ROUTLEDGE LIBRARY EDITIONS: LOGIC

Volume 23

LOGIC IN PRACTICE

T0203797

LOGIC IN PRACTICE

L. SUSAN STEBBING

Routledge
Taylor & Francis Group

LONDON AND NEW YORK

First published in 1934 by Methuen & Co. Ltd
Fourth edition published 1954

This edition first published in 2020
by Routledge
2 Park Square, Milton Park, Abingdon, Oxon OX14 4RN

and by Routledge
52 Vanderbilt Avenue, New York, NY 10017

*Routledge is an imprint of the Taylor & Francis Group, an
informa business*

British Library Cataloguing in Publication Data
A catalogue record for this book is available from the British
Library

ISBN: 978-0-367-41707-9 (Set)
ISBN: 978-0-367-81582-0 (Set) (ebk)
ISBN: 978-0-367-42264-6 (Volume 23) (hbk)
ISBN: 978-0-367-42630-9 (Volume 23) (pbk)
ISBN: 978-0-367-85436-2 (Volume 23) (ebk)

Publisher's Note
The publisher has gone to great lengths to ensure the quality
of this reprint but points out that some imperfections in the
original copies may be apparent.

Disclaimer
The publisher has made every effort to trace copyright holders
and would welcome correspondence from those they have been
unable to trace.

Logic in Practice

L. SUSAN STEBBING
D.LIT., M.A.

*Late Professor of Philosophy in
the University of London*

METHUEN & CO. LTD, LONDON
36 Essex Street, Strand, WC2

First published January 18th 1934
Second Edition, February 1946
Third Edition, May 1949
Fourth Edition, revised and reset, 1954
Reprinted 1959

4.2

CATALOGUE NO. 4623/U

PRINTED IN GREAT BRITAIN
BY JARROLD AND SONS LIMITED, NORWICH

TO
MY SISTERS

'No book can do ALL a man's thinking for him. The utility of any statement is limited by the willingness of the receiver to think.'—EZRA POUND

PREFACE

WHAT passes for knowledge in ordinary life is most often nothing but beliefs which we hold more or less tenaciously without any clear awareness as to what precisely we are claiming to know. Even when our beliefs happen to be true, as is sometimes the case, our lack of precision and our ignorance of the grounds upon which these beliefs could be based permit us to hold other beliefs which are contradictory. We are apt to be cocksure where we should be hesitant, to be vague where precision is important, and to be contentious although argument is possible. These defects are obvious in the case of other people's assertions; we must be exceptionally fortunate—or unusually stupid—if we have never noticed them in our own thinking. It must be the desire of every reasonable person to know how to justify a contention which is of sufficient importance to be seriously questioned. The explicit formulation of the principles of sound reasoning is the concern of Logic.

The study of logic does not in itself suffice to enable us to reason correctly, still less to think clearly where our passionate beliefs are concerned. Thinking is an activity of the whole personality. Given, however, a desire to be reasonable, then a knowledge of the conditions to which all sound thinking must conform will enable us to avoid certain mistakes into which we are prone to fall. There is such a thing as a habit of sound reasoning. This habit may be acquired by consciously attending to the logical principles of sound reasoning, in order to apply them to test the soundness of particular arguments. No doubt there are a few gifted persons whose critical temper of mind enables them to reason soundly although they have never had occasion to attend to the principles in accordance with which their reasoning proceeds. There may be others too stupid ever to be able to appreciate the logical force of an argument. Most people, however, are between these extremes. Their reasons are sometimes sound, sometimes unsound, but they often do not know why they are the one or the other. It is for such people that this book is intended.

It has been impossible in so small a book to do more than touch upon many topics which are worth detailed consideration. Technicalities have been deliberately avoided, for this book is in no sense intended to provide an introduction to Logic. Stress is laid upon the importance of considering language, which is an instrument of our thinking and is imperfect, as are all human creations.

I desire to thank Mr. A. F. Dawn and Miss J. Wynn Reeves for their help in the correction of the proofs, and I am further indebted to the latter for her help in the compilation of the index.

L. S. S.

BEDFORD COLLEGE
UNIVERSITY OF LONDON
December 1933

NOTE TO FOURTH EDITION

The text has been thoroughly revised and corrected for this edition by C. W. K. Mundle, Lecturer in Philosophy at University College, Dundee.

CONTENTS

CHAPTER I

PURPOSIVE THINKING

'Where the senses fail us reason must step in.'—GALILEO

THINKING is an activity; we think in order to do. But not all *doing* consists in overt action producing perceptible changes in the given situation or environment. The 'man of action' is commonly opposed to the 'man of thought'. There are good grounds for this opposition; but even men of action have to think, however much their activities may suggest the contrary. The world today needs clear thinkers even more than it needs men of good will, and not less than it needs men of great practical energy. To be confronted with a problem is to be compelled to think. Thinking essentially consists in asking questions and attempting to answer them. To ask a question is to be conscious of a problem; to answer correctly is to have discovered its solution. Purposive thinking is thinking directed to answering a question held steadily in view. Such directed thinking may be contrasted with idle reverie.[1]

Suppose a man lying awake in his cabin on board a passenger steamer. He listens to the sound of the sea, to the numerous slight sounds—the creakings and strainings—always audible on board ship. His hearing of these sounds may partly determine the flow of his thoughts; he passes idly from one thought to another. Suddenly he hears a loud, distinctive sound—the three long booms which are the danger-signal. *This* sense-impression is significant; he does not notice it merely *as a sound*; it signifies for him—*ship in danger*. He springs up, snatches a coat, and rushing out hears the word 'Fire!' The reader's imagination may supply the details. Provided that the man be not too panic-stricken to think at all, his thinking will now be purposive; it will be

[1] Cf. C. A. Mace: *The Psychology of Study*, Chap. IV. In this chapter will be found an excellent account of the process of thinking, treated from the point of view of the psychologist.

1

directed to securing his own safety or that of others. He will now actively connect one apprehended fact with another. Once the *fire-situation* is grasped, his thinking will be directed to a practical end; the conditions of attaining this practical end will constitute the problem which his thinking is directed to solving.

Suppose now that a committee of investigation is confronted with the problem of how the fire originated. This problem is purely theoretical, however much the desire to solve it may be the practical desire of assessing the responsibility for the outbreak of fire, or of attempting to prevent the occurrence of such accidents in the future. A problem is not made a practical problem simply because its solution may have practical applications. The committee are seeking to obtain knowledge; they want to find a true answer to a definite question. Their problem is as purely theoretical as the question of determining the conditions of combustion in general, or the problem of determining the nature of eclipses. The distinction between what is often called *practical* thinking and *theoretical* thinking lies wholly in the purpose for which the thinking is pursued. In both cases the thinking process is the same; it is purposive, and thus directed. The opposition is not between practical and theoretical thinking but between directed thinking and idle reverie.

Intelligent dealing with a problem involves, first, the apprehension of a situation giving rise to the problem; secondly, the explicit awareness of the question constituting the initial stage of the problem; thirdly, formulation of the conditions to which the solution must conform. These conditions are determined by the total situation. In so far as these conditions are clearly apprehended and selectively attended to, precise questions can be formulated and various answers tried out. The point to be stressed is that to ask an intelligent question is to have taken note of the conditions set by the problem; to suggest an intelligent answer is to have discerned within the situation, so far as apprehended, those factors which may be relevant to the solution. Intelligent answers may be wrong, but they are never off the point. To keep to the point is to be guided by relevant considerations alone.

The importance of excluding irrelevant suggestions cannot be overstressed. In the case of the man on the burning ship, it is clear that his thinking would be effective only in so far as what came into his mind—or, at least, was taken note of—was relevant to the conditions constituting his problem. If he were to consider whether he could fly like a bird from the ship, or whether the flames could be extinguished by a heavy rainfall, he would be asking questions the answers to which could have no bearing upon the difficulty with which he was confronted. In the same way the committee investigating the cause of the fire would be making no progress towards the solution of their problem if they were to ask whether the fire was due to the ship's having sailed from port on a Friday, or whether it was caused by a hot dispute between members of the crew, or whether it was due to the wrath of God because the passengers were dancing on a Sunday. That these suggestions at once strike the reader of this book as absurd is due to his knowing too much about the kind of situation, and thus about the conditions of the problem under discussion, even to entertain the possibility that such factors could be relevant. It might well be relevant, however, for the committee to ask whether the fire was due to the careless dropping of a lighted match or of an unextinguished cigarette end, or to a faulty electrical cable, or to a deliberate act of incendiarism. Each of these questions would suggest other questions the answers to which might more easily be ascertained to be correct or incorrect. In this way progress towards a solution may be made. We may briefly consider each of the last three suggestions in order to see how a relevant answer may admit of testing.

The question the committee sought to answer was quite definite, namely, What caused *this* fire? They sought to discover, not *a possible* cause of fire on board ship, but *the actual* cause of *this fire*. Each suggestion presents a possible cause. Other possible causes might have been considered. The only way to test these suggestions is to ask *what else* would have happened if the given suggestion were correct, but would probably not have happened if one of the other suggestions were correct. To be able to ask these further questions the committee would need to know a good deal about the ship and about what took place when the fire occurred; they would also have to possess certain technical knowledge. If the ship had been completely burnt out, some relevant questions could not be answered. If, however, the burning ship had been towed into port or there were survivors who could answer definite questions, and if the committee possessed *other* relevant knowledge, it might be possible to say what was most probably the cause. These conditions may be assumed to be fulfilled, for without some means of obtaining such information there would be nothing to investigate.

Each of the suggestions put forward is a supposition to the effect that some occurrence happened, this occurrence being such that, if it *had* happened, then fire *would* have broken out. A supposition thus entertained in order to account for what happens is called an *hypothesis*. The committee seriously considered *three* hypotheses. Each hypothesis has certain *consequences*, i.e. given that the hypothesis were correct something else would have happened. The question then arises whether it did happen.

(1) The hypothesis that the fire was caused by a carelessly dropped lighted match suggests the following questions: (*a*) Did the fire begin in a cabin or in a public part of the deck? (*b*) Did it occur at night? If the answer to (*b*) is affirmative, and if the answer to (*a*) is that the fire broke out in the luggage-room, then it is improbable that this first hypothesis is correct, since it is unlikely that any one would be smoking in the luggage-room, especially at night.

(2) The hypothesis that the cause was a fault in an electrical cable suggests the questions: (*c*) Did the fire originate in the vicinity of an electrical cable? (*d*) Had the electrical installation in this part of the ship been recently overhauled so that it was known to be in good repair? (*e*) Did anyone notice, at the time when the fire broke out, that in some parts of the ship the electrical supply was cut off, indicating a blown fuse? An affirmative answer to (*c*) would not of

course rule out the lighted match hypothesis, but if the fire had broken out at night and in a little frequented part of the ship, then the fact that it originated in the vicinity of an electric cable would tell in favour of the present hypothesis. But if the answer to (d) were 'Yes' and to (e) were 'No', then the available evidence would not be in favour of this hypothesis. At this stage the third hypothesis seems plausible, namely, that some one deliberately set the ship on fire.

(3) This hypothesis is not one which would readily be accepted. Careless smokers and electrical faults are familiar enough. Certainly incendiarists are not unknown. But no sane, ordinary passenger would put himself to the inconvenience, perhaps even peril, of being on a burning ship. Thus, unless the incendiarist were a madman, he must have had some strong motive for so acting. This hypothesis suggests the question whether any one on board could have expected to derive some benefit from the destruction of the ship, or was the agent of some one who had such expectations. In attempting to answer this question the committee would be led to pursue investigations beyond the occurrences on the voyage itself. They would ask *what* benefit could be derived from the destruction of the ship, and *who* would be benefited. Further questions at once arise. (*f*) Was the ship heavily insured? (*g*) What was the age of the ship? (*h*) Were the owners in need of money? Suppose it were found that the ship was heavily insured; that its future sea-going life was not likely to be long; that the owners needed the insurance money; then this third hypothesis is worth taking seriously. If it were further discovered that other ships belonging to the same Line had been recently destroyed by fire, then the hypothesis will seem not unreasonable under the circumstances.

It is not worth while to pursue this illustration further. Enough has been said to show how an intelligent person confronted with a problem will proceed to ask questions and guess at the answer; how various answers lead to other questions and further guesses. A guess is worth making only if the answer can be tested. These guesses are suggestions

as to what may possibly be the case; i.e. they are hypotheses. These hypotheses are worth entertaining only if
the possibilities can be narrowed down. The possibilities
can be narrowed down by discovering that what would be
possible in certain other circumstances is not possible under
the ascertained circumstances. In the next chapter we shall
see how such an investigation conforms to certain principles
which interest the practical logician.

Thinking out a problem involves inference. Inference is
a mental process in which a thinker passes from something
given or taken for granted—the datum—to something which
he accepts because, and only because, he has accepted the
datum. It is a passage of thought from datum to conclusion.
To accept a conclusion as the result of an inference is to
accept it upon the basis of what is taken to be evidence. To
regard what is observed, or is believed, or is in any way
apprehended, as *evidence* is to regard it as indicating something other than itself. To regard a fact as providing evidence is, then, to regard that fact as significant of some other
fact. We may have *some* evidence in favour of a conclusion
and no evidence contrary to it, and yet the evidence may not
be *decisive*. Evidence is decisive for a certain conclusion
when the acceptance of the evidence necessitates the acceptance of the conclusion. Unfortunately we may be mistaken
both with regard to what the facts are and with regard to
what the facts indicate.

The example of the committee's investigation has shown
us that intelligent guessing is controlled by the recognition
of certain ascertained conditions as relevant to the solution
of the problem. These conditions relate to matters of fact.
They may be called the material conditions of the problem
since they are provided by the special subject-matter of the
investigation. To apprehend a condition as relevant to the
solution is already to know a certain amount about the
nature of situations resembling the given one in certain
respects. If we knew *all* about the situation there would be
no problem. If we knew *nothing* about *similar* situations we
should not be able even to ask intelligent questions about it.
Relevance is significance for the question at issue. *Nothing*

is significant in itself. That a ship is heavily insured *acquires* significance if the ship belongs to owners in financial difficulties and if its destruction occurred in a manner compatible with a deliberate act of incendiarism; it *ceases* to be significant if the owners are known to be men of incorruptible honesty. A red light at a street corner signifies '*stop*' to a motorist who knows the conventions of traffic regulation. A certain-shaped mark on the sand is significant of the previous presence of a human being only to one who can interpret the mark as a *footprint*. A grey mark on a carpet may be significant of a certain brand of cigars to one who, like Sherlock Holmes, has made a special study of varieties of cigar ash. To multiply examples is unnecessary. The point must be obvious to any one on reflection. Nevertheless, we are apt to forget that significance arises only in so far as a given fact indicates something other than itself. The possibility of such indication depends upon knowledge possessed by the person apprehending the significance. The detective who 'takes in the whole situation at a glance' would need already to know *what* each item he observes *signifies*. In sober life this is not the case. The observed facts acquire significance only when viewed in the light of a definite question which guides his thinking. Reviewed in the light of a different question, the same facts may alter in significance. Readers familiar with detective novels will be able to supply examples illustrating this point.

It should now be clear that *significance* is a property of *signs*. A sign indicates something other than itself. Signifying is a relation requiring three terms: a sign, that of which the sign is significant, and a thinker for whom the former indicates the latter. Just as a book, for example, is not a *gift* unless it be given *by* some one *to* some one, so a red light, for example, is not a *sign* unless it be *interpreted by* some one to *indicate* something. To be able thus to interpret a presented fact we must know something not given in the presented situation.

It is a mistake to regard thinking which involves interpretation of signs as confined to the higher levels of consciousness, or to confine 'problems' to what are often called *intellectual* problems. There are no hard and fast distinctions in the development of mental life. The distinction between practical and theoretical problems does not relate to the kind of thinking involved, but to the set of purposes within which the problem originates, and hence to the kind of changes its solution is designed to bring about. A problem may be practical in the sense that it relates to bringing about a change in the environment, for example, making a box, or mending a clock, or disposing an army, or winning an election campaign. A problem may be theoretical in that it relates to bringing about a change in the mind of a thinker who has no other purpose in view than to apprehend a situation more clearly by discerning the connexion between its component elements. The purpose of solving the problem is, in the latter case, the acquirement of knowledge; in the former, the doing of something. In both alike, thinking is directed to an end determined by the nature of the problem. In effective thinking consciousness of the conditions set by the problem-situation directs the cognitive activity of the thinker, determining what shall come to mind.

Directed thinking in its most highly developed form is reasoning. To be reasonable is to be capable of apprehending a situation as a whole, to take note of those items which are relevantly connected, and not to connect arbitrarily items not apprehended as significant. Apprehension of relevance depends upon two quite different factors: knowledge and sagacity. 'To be sagacious', William James has said, 'is to be a good observer.' This statement may be accepted if we admit that a good observer is one capable of discerning relevant connexions. Certainly a good observer is not one who 'stands and stares'. Staring is a sign of stupidity, or of an astonishment so profound as temporarily to destroy the power to think reflectively. In reasoning we select from a set of items presented to us just those which are significant in relation to the facts we are trying to establish. Failure to be reasonable may occur in one or other of two very different ways. Significance may be missed through sheer ignorance or through passion and emotional attitudes which check clear-sighted apprehension. No doubt we all desire to be reasonable but few of us achieve this desire. Even those few are reasonable only at rare moments. Ignorance and passion present grave obstacles to be overcome only by a supreme desire to think clearly. The scientist is regarded as the exemplar of rationality. Within the field of his expert investigations he is, for here he has both relevant knowledge and a disinterested desire to understand merely for the sake of understanding. Confronted with problems of a different kind, such as those presented by a political crisis or by the behaviour of his children, he may allow passion to subdue reasoning, stupidly asking irrelevant questions and accepting irrelevant answers.

Correct answers are as rare as they are difficult. Human beings from infancy upwards are confronted with problems. Our ability to deal with these problems effectively depends in part upon our ability to think clearly. It is as natural for men to think as to walk and to speak. Few of us, however, walk or speak well, although we may remain for ever unaware of our deficiencies in this respect. Where few attain a high level of excellence, the majority are apt not to notice that their performance falls short. Our natural capacity to walk or to speak may be improved by practice, guided by some standard which we can try to reach. The case is not otherwise with thinking. Although we all must think, we seldom think effectively. Our thinking is more likely to be effective if we are aware of the conditions to which efficient thinking must conform. To know these conditions is to have a standard by reference to which we may gauge the success of our efforts. In this way we may learn to avoid some mistakes.

CHAPTER II

THE IMPORTANCE OF FORM

'Importance depends on purpose and on point of view.'
A. N. WHITEHEAD

THE problems about which we can think effectively arise out of situations which are on the whole familiar: Were this not so we should not know what questions to ask, still less what answers were relevant. In this book we are not concerned to inquire how we come to have such familiarity. We are not asking how knowledge originates. Our problem is limited to considering how we may reasonably come to accept conclusions we did not know before and now accept because, and only because, we have accepted something else. Knowledge thus obtained is obtained by inference. It is not always easy to draw the line between a judgment in which no inference is involved and an inferred conclusion. In judging 'That is a cow', 'Here is a pen', 'That is a motor passing down the street', we are merely recording what we take to be directly given. It is true that such judgments go beyond what is directly given to sense, but it does not follow that they are reached by inference. If they are questioned we may attempt to justify them by inference from something indubitably given. Certainly such judgments may be erroneous. Nevertheless, recognition, perception, memory, provide us with the materials from which we start. None of these is completely trustworthy, but they are all we have.

The passenger on the ship—in the illustration considered in the last chapter—*recognized* a certain sound as a *danger-signal*. Had he not so recognized the sound, there would have been no problem for him. The committee of investigation *assumed* that happenings do not occur haphazardly, but are so connected that what happens *here—now* is conditioned by what happened *there—then*. In their inquiry they took for granted a number of generalizations with regard to the way things of a certain sort behave in relation to other things, e.g. lighted matches in relation to wood, careless smokers in relation to lighted matches, fuses in relation to electric cables. *Things of a certain sort* constitute a *class*. The phrase 'of a certain sort' indicates that the things possess in common, properties which do not all belong to any other thing. Class-names, or common nouns, are used to refer to all the things of a certain sort. The class-name 'ship', for example, is used to enable us to refer to many objects each of which behaves in the same sort of way in relation to *other* things. No two ships have all their properties in common, but the differences between one ship and another can often be ignored as *irrelevant* to what we are thinking about. We may, however, wish to take note of some of these differences. If we are in a hurry to cross the Atlantic Ocean it matters to us whether the ship is propelled by steam or by wind, whether it has efficient engines, and so on. Thus we distinguish between 'steam ships' and 'sailing ships', between 'screw steamers' and 'turbine steamers', between 'sloops', 'schooners', 'cutters', and 'brigs'. A ship can also be regarded from the point of view of its tonnage, or of its capacity for holding so many passengers, or of its belonging to a certain Line, or of its being heavily insured. In taking note of these different properties we put a ship into different classes. A given definite object is a member of many different classes. A class is nothing but a set of objects possessing in common properties not all of which are possessed by any object not belonging to that class. Whenever the common possession of these properties is important for our purposes we use a class-name. When we want to distinguish from other objects a set of objects possessing in common a property not previously noted, we invent a class-name, e.g. 'X-rays', 'Bolshevik', 'taxi'.

The use of class-names enable us to economize thought. It would be inconvenient if we could not briefly refer, e.g., to all those vehicles which we now distinguish from other vehicles as 'horse-cabs'. What we, in England, understand by 'taxicab', and its familiar abbreviation, 'taxi', is not distinguished merely by the possession of a taximeter, so that the Englishman in Italy may be surprised to find a horse-drawn vehicle answer his summons for 'a taxi'. *How* we use

12   LOGIC IN PRACTICE

a word, i.e. what we mean to refer to by using it, is largely
determined by the situations in which we have met the
objects to which the word refers. We shall see later that
here, in the use of familiar class-names, lies a certain danger
to clear thinking.[1] At present we are concerned only to
notice how indispensable is this convenience of language.
Class-names enable us to abbreviate and to connect. The
psychological reasons which make the invention of such
words as 'taxicab' and 'Bolshevik' useful also lead to the
specially devised terminology of a special science. The
chemist finds it useful to speak of 'carbon monoxide' and
'carbon dioxide', and even to abbreviate further by using the
carefully devised symbolism CO and CO_2. Formerly
(about 1630) carbon dioxide was called '*gas sylvestre*'; later
(about 1754) it was called 'fixed air'. Each of these names
is significant; the former of its discoverer, the latter of one
of its properties. The name 'carbon dioxide' is, however,
more importantly significantly to the chemist in showing
the way in which this gas is composed. The reader will be
able to think of many other examples of such convenient and
economical procedure. Words used in ordinary situations
are not so deliberately devised. They are used to indicate
the presence of characteristics, or properties, which we have
frequently found to be possessed by an object in various
situations. The point to be stressed is that in using the
common nouns which belong to our everyday vocabulary we
are benefiting by knowledge acquired throughout the course of
human history. A class-name refers to *important* properties.

Throughout the preceding discussion the word 'important' has been frequently used, and the corresponding
abstract noun appears in the title of this chapter. It is
desirable to understand what the word 'important' means.
To say of anything that it is 'important' is to say that it 'makes
a difference relevant to our purpose'. Nothing has importance apart from a purposive being. What makes a difference for one purpose makes no difference for another
purpose. Hence, importance is relative to a point of view.
It must be insisted that for different purposes different

[1] See Chap. IV.

properties are important. The reader would be well advised if, whenever he meets the word 'important' in a serious discussion, he were to ask himself: 'Relative to *what purpose*, or from *what* point of view, is this important?' A few examples may make this point clear.

Men are alike in certain respects; they differ in others. Similarly, with *cows, ships, rainbows, novelists, conservatives,* or any other familiar class the reader can think of. Although *men* differ from *cows* in certain respects they are like them in others. The respects in which men are alike are *important*; hence we have the class-name. The differences between a *man* and a *cow* are important; we therefore refer to them by different class-names. But since men and cows are alike in that both are animals, we have another class-name, 'animal', connecting them into a wider class. Animals differ in important respects from ships; they are like them in having weight and being movable. 'Important' has here been used to indicate 'making a difference to the ordinary purposes of ordinary men'. So important for these purposes are the differences between the classes mentioned, that the grouping together of the objects in each class, and the distinction of the class thus formed from other classes, forms part of our everyday knowledge. The various likenesses and differences are *obvious* in any situation; they cannot fail to strike our attention. Sometimes, however, an unobvious likeness may be more important than a striking difference. A good example is given by Bain, who points out that 'we become oblivious of the difference between a horse, a steam engine, and a waterfall, when our minds are engrossed with the one circumstance of moving power'.[1] To single out the characteristic of *moving power* involves the imaginative selection of one characteristic out of a complex situation in order that it may be attended to in isolation. In so doing we ignore differences which, from another point of view, would be important. Thus we analyse a situation and abstract characteristics discernible within it. This process of analysis and abstraction is involved in finding a hidden face in a puzzle-picture.

[1] *The Senses and the Intellect*, p. 521. Cf. L. S. Stebbing: *A Modern Introduction to Logic*, pp. 6–7.

On the basis of past experience what is first apprehended as a medley of tangled lines may come to be seen as a sketch of a man smoking.[1] In such apprehension abstraction, analysis, and subsequent synthesis are involved.

What we apprehend, then, is conditioned by what we have previously apprehended. This should have been made clear by our discussion of significance in the preceding chapter. For the purposes of inference an *important* property is a property which can be taken as indicating the presence of another property. In our everyday experiences we do find properties constantly so conjoined that we can infer from the presence of one to the presence of another. Since things of the same sort exhibit characteristic modes of behaviour in determinate situations, the recognition of an object as belonging to a certain class may enable us to discover how it will behave, or how it has behaved, on some unobserved occasion. We say *Wood burns easily*, meaning thereby to assert that *every* piece of wood, will so behave in the presence of fire; we are not referring only to those pieces of wood which have been observed to burn. This character-istic mode of behaviour on the part of wood *is* its property of being *inflammable*. If we say *The broadcast speeches of politicians do not express the views they really hold*, we are saying something about the way politicians behave in the situation of broadcasting speeches. Such statements are made about the whole of a class on the ground of the observed characteristics of some of its members. A statement of this kind is an empirical generalization. The use of a class-name itself results from generalization, since, in applying the class-name to an object we are asserting that the object possesses properties not directly observed and belonging also to other objects, although we may not be aware that we are doing so until the application of the class-name is challenged. Empirical generalizations may be false; class-names may be wrongly applied. As practical logicians we need to ask under what conditions we may reasonably rely upon empirical generalizations and may safely use

[1] Cf. A. W. P. Wolters: *The Evidence of our Senses* (Methuen's Monographs on Philosophy and Psychology), p. 41.

class-names. It is the purpose of this book to afford some help towards answering these questions.

The kind of inference involved in *reaching* generalizations must be distinguished from the kind of inference exemplified in applying generalizations to particular cases. In this chapter we shall be mainly concerned with the latter, but a few words may be said about the former kind of inference.

Generalization involves a general property, i.e. a property which may belong to many things. Having noticed that all the things of a certain sort, which have been observed, behave in certain ways, we infer that *any other* thing of that sort, although it has not been observed, will behave in that way. For example, if, on the ground that all the psychoanalysts we have met have been deficient in humour, we conclude that every psychoanalyst lacks humour, we are generalizing from *some* members of a class to *all* its members. This is an example of what is known as inductive inference. The observed instances constitute the datum of the inference. The datum provides the premiss of the inferred conclusion. We should not claim that those psychoanalysts we happen to have met are all the ones there are; indeed, we want to assert that deficiency in humour is *characteristic* of them *as a class*, containing observed and unobserved members. It may well be that the ones we have met were an unfortunate selection; they may not have been representative. A single exception contradicts the generalization. It is the characteristic of inductive inference that the premisses may be true and yet the conclusion may be false. This is possible because inductive inference *goes beyond* the evidence.

Let us suppose that the application of a class-name has been challenged. How do we proceed to answer the challenge? We seek to point to some characteristic which belongs to every member of that class and to nothing else. Let us imagine two people looking out to sea. A says, 'That's a cutter.' B replies, 'No; it is a sloop.' Here A's application of the class-name 'cutter' has been challenged. In replying to the challenge A will make use of previously acquired knowledge. He may answer as follows: 'It *is* a cutter, for not only is it single-masted, but it has a running bowsprit and no jib-stay.' A has now presented his original statement as *the conclusion* of an inference; he has supported the statement by producing *reasons*. In thus reasoning A has made use of previous knowledge, involving generalizations, to suggest characteristics the possession of which affords a test of the correctness of the original statement.

Since we are not interested in the mythical dispute between A and B, but in the nature of A's reasoning, we may proceed to set out the steps of his reasoning at a length which people engaged in the pleasant pastime of watching ships coming into harbour could only regard as tedious and pedantic. The steps may be set forth as follows:

> *Alternatives:* It is a cutter or a sloop.
> *Suggested test:* Has it a running bowsprit and a jib-stay?
> *Argument:* All sloops have standing bowsprits and jib-stays;
> This ship has a running bowsprit and no jib-stay;
> ∴ This ship is not a sloop.
> All cutters have running bowsprits and no jib-stay
> This ship has a running bowsprit and no jib-stay;
> ∴ This ship is perhaps a cutter.
> But (it was agreed) it is either a cutter or a sloop;
> And (it has been shown) it is not a sloop;
> ∴ This ship is a cutter.

This reasoning is an example of *deductive* inference. The *reasons* offered are the *premisses* of the inference. These premisses are taken to be true, and it is shown that, this being so, the conclusion *must* be accepted. To say that we *must* accept a conclusion is to say that we should not be rational in accepting the premisses and rejecting the conclusion. This is the distinguishing characteristic of deduction. It would be quite reasonable to accept the premiss, *Some psychoanalysts lack humour* whilst admitting that possibly not all do. The formal distinction between inductive and deductive inference consists in the fact that the conclusion of an inductive inference may be false although the premisses are true; whereas, the conclusion of a correct deductive inference cannot be false provided that the premisses are true. Hence, in deduction the truth of the premisses is a guarantee of the truth of the conclusion.

It should be observed that we have said '*provided that* the premisses are true'. The truth or falsity of the premisses is determined by their relation to facts, i.e. to definite states of affairs which are the case. This relation of premiss to fact is what, in Chapter I, we called the *material conditions* of a problem. It may be the case that it is false that all sloops have standing bowsprits and jib-stays; in the argument, however, it was taken for granted that the premiss was true. We cannot (as we shall see later) establish *its* truth by deductive inference. We can only assert that *if* it is true, *then* something else *must also* be true. The 'must' expresses the *formal condition*. The study of formal conditions is the special business of the logician. It is the purpose of this chapter to make clear the nature of formal conditions.

In the preceding discussion the words 'true' and 'false' have been frequently used. It is to be assumed that the reader knows quite well how to use these words.[1] It is sufficient for our purposes to notice that whatever can significantly be said to be true, or false, is a *proposition*. The answer to a question is always a proposition. Whenever a person makes a statement he is putting forward a proposition as true. Commands, requests, prayers, and questions are not propositions. Of none of these could truth or falsity be significantly asserted. Using the notions of truth and falsity we can define the relation upon which deductive inference depends. This is the relation of *implication*, or—as we shall often call it—*entailing*. A given proposition *entails* another proposition when there is between them such a relation that the truth of the given proposition is inconsistent with the falsity of the other proposition. This relation of *entailing* holds between the premisses, taken together, and the conclusion in the argument about the cutter, given on page 16. The premisses *entail* the conclusion; the

---

[1] The determination of what 'true' and 'false' respectively mean is a philosophical problem, which lies outside the scope of this book.

conclusion *follows from* the premisses. The relation of entailing
is very important when we want to make use of knowledge
we already possess in order to discover something we did not
know. If we can find a proposition which entails another,
and if we know that the entailing proposition is true, then we
know that the entailed proposition is true. Suppose we
know that *all cutters are single-masted*. We can see at once
that *whatever is not single-masted is not a cutter*. Each of these
propositions entails the other. It would thus be irrational
to accept one and reject the other. In the case of this simple
example no one would be likely to do so. Indeed, the reader
may feel that to say 'whatever is not single-masted is not a
cutter' is only a more awkward way of saying 'all cutters are
single-masted'. Let us consider another example: *If the fire
was caused by a fused wire, then it would spread along the
electric cables.* This proposition entails the proposition: *If the
fire did not spread along the electric cables, then it was not
caused by a fused wire.* Each of the italicized statements is a
single proposition. Either entails the other. The fact that
the one entails the other is quite independent of its truth or
falsity. We may be mistaken in supposing that there is such
a connexion between the fusing of a wire and the spread of
the fire along the cables; whether, or not, there is such a
connexion depends upon material conditions relating to the
behaviour of *fused wires, fire, electric cables.* The ascertain-
ment of these conditions requires special knowledge concerning matters of fact. But the truth of the statement that
one of the above propositions entails the other is quite
independent of matters of fact; the statement is to the effect
that a *formal* relation holds between the two propositions
whether they are *both true* or whether they are *both false.*
This formal relation is such that we *must* accept both or
reject both, since either entails the other.

One proposition may entail a second although the second
does not entail the first. For example, *All sailors are
superstitious* entails *Some sailors are superstitious,* but not
conversely. Thus the relation of entailing is not simply
reversible, although two propositions may entail each other.
As soon as we understand the nature of entailing, we can

formulate a fundamental logical principle, namely, *Whatever is entailed by a true proposition is true.* We shall call this the *Principle of Deduction*, for it is in virtue of this Principle that we can validly infer one proposition from another. This Principle lies at the basis of all conclusive reasoning. The reader may never have met the Principle in this abstract form, yet he will often have reasoned in accordance with it. We apply the Principle whenever we argue that a given conclusion *must* be accepted *because* certain premisses have been accepted. We may, of course, be mistaken in supposing that the premisses are thus related to the conclusion. Various ways in which we are prone to make this mistake will be mentioned in the next two chapters. Here it is sufficient to notice that this Principle is a *formal* condition of deductive inference.

Let us go back to the committee of investigation whose deliberations were sketched in the last chapter. We have now to notice that their thinking was controlled by formal, no less than by material, conditions. No doubt they were not explicitly aware of these formal conditions, but their thinking was effective only in so far as it was in accordance with them. The committee would have been simply stupid if, having guessed that the fire was caused by a lighted match, and having admitted that, *in that case*, the fire would almost certainly have broken out either in a cabin or in a public part of the ship, they had nevertheless stuck to this guess *although* it had been ascertained that the fire broke out in the luggage-room, i.e. neither in a cabin nor in a public part of the ship.[1] The way in which their thinking was controlled by formal conditions may be clearly shown if the steps of their reasoning are set out at the tedious length required to exhibit all the conditions determining the direction of their thinking. This we can do in the following manner:

Problem: Something, we don't know what, happened, AND THEN, fire occurred. (An observed fact.)

[1] The reader should consider this long statement, in order to see *why*, if the committee had so reasoned, they would have been stupid.

Question: What happened?

First Guess: An unnoticed lighted match came into contact with an inflammable part of the ship, and set fire to it.

I. *Testing the Guess:*

    (1) If so, then the match was dropped in a cabin or in a public part of the ship, and the fire began in the place where the match was dropped.

    (2) But, the fire broke out in the luggage-room (i.e. not in a cabin nor in a public part of the ship).

    (3) Therefore, the cause of the fire was not a lighted match.

Second Guess: The fire was caused by a short-circuit in one of a certain number of electrical cables.

II. *Testing the guess:*

    (1) If so, one or more of certain fuses would have been blown.

    (2) But none of these fuses was blown.

    (3) Therefore, the cause of the fire was not such a short-circuit.

Third Guess: Some one deliberately set the ship on fire.

Further Question: Who could want to set a ship on fire?

Tentative Answer: Some one who would benefit by its destruction.

Further Question: Who would benefit in this case?

III. *Testing the Guess (in the light of the further questions):*

    (1) If the ship were deliberately set on fire, some one would benefit by its destruction.

    (2) If the ship were over-insured, the owners would benefit by its destruction.

    (3) But the ship is over-insured.

    (4) Therefore, the owners benefit by its destruction.

What exactly has this reasoning established, and how has it established it? It must be observed that, so far as we have gone, it has *not* been shown what the cause was, nor that

the owners had anything, directly or indirectly, to do with
the burning of the ship. All that has been shown is that
certain *possible* causes of fire on board ship were not, given
certain assumptions, the *actual* cause of this particular fire.
We have now to examine the way in which the reasoning
proceeded. It should be noticed that the committee, faced
with the problem, did not immediately obtain premisses
which entailed the answer to their question. On the contrary, they had to *jump* to a possible conclusion, and then
test its correctness. Each successive guess led to the formulation of an hypothesis, regarded as having certain consequences. The reader will see that the reasoning in I and
II proceeds in precisely the same way. In each case the
hypothesis is rejected because the consequence was found
not to be the case. This reasoning may be schematically
represented, in a shortened form, if we use H_1, H_2 to stand
respectively for the first two hypotheses, and C_1, C_2 for
their corresponding consequences. The scheme is:

(I) If H_1, then C_1,   (II) If H_2, then C_2,
    but not C_1,   *but* not C_2,
    $\therefore$ not H_1. $\therefore$ not H_2.

It is easy to see that in both cases the reasoning is in conformity with the Principle of Deduction, *whatever is entailed
by a true proposition is true*. This principle tells us that we
must not accept a given proposition, H, and reject a proposition, C, which is entailed by H. (To reject C is to say
that C is false.) We cannot accept a hypothesis and reject
its consequences. It is by finding that its consequences are
false that we can eliminate an hypothesis.

The reasoning in I and II is said to have the same form.
The reasoning in III is of a different form, which will be
discussed later.[1] The reader will have had no difficulty in
recognizing the soundness of the reasoning in each case. It
is easier to see that an argument is sound, or unsound,
than to see wherein its soundness, or unsoundness, consists.
But to have insight into the conditions of sound reasoning
is very important for us as practical logicians.

[1] See p. 69 below.

3

It must be noticed that in discussing the conditions upon which the soundness of an argument depends we do not need to consider a particular argument. We need not have taken the example of the ship on fire; we might have discussed the authorship of the *Book of Job*. What we said about the reasoning in I and II was quite general; it related to a *form* of reasoning. Very many arguments could be fitted into this form. The *conclusiveness* of an argument depends entirely upon its form. Sound reasoning is *valid* reasoning.

The validity of reasoning depends upon purely formal conditions. These conditions are quite general, and are thus formal; hence they are independent of special matters of fact.

All reasoning, when fully stated, has a formal aspect. This does not mean to say that all reasoning is *deductive*, although all *conclusive* reasoning is deductive. It means that if our reasons are sound reasons in a given case, they must be sound in the case of any other argument which has the same form. The notion of *the form of an argument* is not familiar to most people. It is an abstraction. A quite simple example may show how the validity of our reasoning depends, not upon the matter-of-fact assertions we are prepared to make, but upon the *form* of the reasoning.

We will suppose that A and B are sitting on a rocky cliff on the Cornish coast. A says, 'There are blasting operations going on.' B says, 'How do you know that?' A replies, 'Because blasting *always* sounds like that.' B says, 'But the sea rushing into the clefts underneath makes a sound like that.' A maintains, 'No. It isn't the sea; it is the sound of blasting.' B objects, 'Well, anyhow, you haven't proved your point. Even if blasting does sound like that, so does the sea when it rushes in underneath the cliff.' If at this stage of the argument A makes the counter-objection that the sound of the sea dashing underneath isn't exactly like the sound of blasting, B may well reply that A's reason was not a *good* reason. It was not a good reason because A should have said, '*Only* blasting makes a sound like that.'

We will suppose that A is obstinate and stupid, and that

B is patient and of a pedagogical turn of mind. A continues to maintain that his original reason was a good reason. Whereupon the following dialogue takes place.

A: 'I don't see any difference between *All blasting sounds like that* and *All that sounds like that is blasting*, except that the second way of putting it is very clumsy.'

B: 'Do you see any difference between *All seals are mammals* and *Only seals are mammals?*'

A: 'Of course. The first is true, and the second isn't.'

B: 'Why do you say the second proposition isn't true?'

A: 'Because men, and horses, and elephants, and a lot of other animals are mammals as well as seals.'

B: 'Then it doesn't follow from *All seals are mammals* that *All mammals are seals?*'[1]

A: 'Of course not.'

B: 'Then you ought to admit that it doesn't follow from the fact that *all blasting operations make a certain sort of sound*, that *whatever makes that certain sort of sound is a blasting operation.*'

It is to be hoped that B's argument may have convinced the reader, whatever may have been the case with A. The point that concerns us is why B began to talk about *seals* and *mammals* in order to show A that his reason for holding that a certain sound was due to blasting was not a *good* reason. There is no connexion between *blasting and sounds*, on the one hand, and *seals and mammals*, on the other. B's purpose was to call A's attention to the *form* of what he said, since, if the reason he offered was a good reason, it must be a good reason in *any other argument of the same form*. Now *all seals are mammals* is related to *all mammals are seals* in the same way as *all blasting operations make that sort of sound* to *whatever makes that sort of sound is a blasting operation*. We can abbreviate the statement if we substitute S for seals and M for mammals. Then we can see that *all S is M* does not entail *all M is S*, nor conversely. In *no* case does *all M is S*

---

[1] Here B assumes that if *what follows* from a given proposition is false, then that given proposition is also false, and A's next remark accepts this assumption.

follow from *all S is M*, no matter what S, or M, represents, and no matter, *therefore*, whether it is in fact the case both that *all S is M*, and that *all M is S*.

We are able to say '*in no case* does it follow', because whether it does follow or not depends, not upon the matter of fact asserted, but wholly upon the form of the assertion. It is for this reason that the logician must insist upon the importance of form, since his purpose is to determine the validity of reasoned arguments. The conditions of validity constitute the formal conditions of a problem.

CHAPTER III

DEDUCTIVE FORMS

'All the inventions that the world contains,
Were not by reason first found out, nor brains;
But pass for theirs who had the luck to light
Upon them by mistake or oversight.'
SAMUEL BUTLER (1612–80)

THE Objection to giving bad reasons is not to be found
in the falsity of the conclusion. On the contrary,
sometimes bad reasons are given in order to support a
conclusion which is in fact true. The objection is that bad
reasons do not *show* that the conclusion is true. 'Bad reasons'
are not properly *reasons* at all, since their badness consists
in their not affording *any* reason why we should accept the
conclusion. Accordingly, if we are shown the unsoundness
of the argument, we shall be left without any ground for
the acceptance of the conclusion. Of course if we firmly
believe that the conclusion is true, we may then look round
for reasons to support it. In this search we may be helped
if we know the kind of premisses which are required to
justify the acceptance of the conclusion. The kind of premisses required will depend upon the form of the argument
into which the premiss has to be fitted.

Reasoning is possible because the truth, or falsity, of one
proposition is not independent of the truth, or falsity, of all
other propositions. Every statement we make has consequences, i.e. implies that other statements are true and still
others false. Most statements we make have grounds, i.e.
are related to other statements which imply them. We often
are not aware of these grounds, nor of these consequences.
In reasoning, however, we seek grounds or we seek consequences. This should be clear from our discussion of
inference in Chapters I and II.

There are seven possible relations which may hold
between any two propositions with regard to the inferability

25

of one from the other. Every one is familiar with these
relations, even if he happens not to know the technical
names which logicians have used for the sake of distinguishing between them.

We will begin by considering the two opposite relations
of *compatibility* and *incompatibility*. Two characteristics are
incompatible when the presence of one necessitates the
absence of the other, and conversely. The following statement illustrates a common use of the word: 'He felt that to
be a politician and a preacher of righteousness was to combine two vocations practically incompatible.'[1] If the reader
thinks that a man may be both a politician and a preacher,
he thinks that these characteristics are *compatible*. Compatible and incompatible have the same significance when
asserted of propositions. One proposition is *incompatible*
with another if they cannot be true together. Propositions
may, however, be *compatible* without being so related that
it is possible to infer the one from the other, or to infer from
the truth, or falsity, of the one to the truth, or falsity, of
the other. The relation of *bare compatiblity* interests no one
except a logician. The proposition *Darwin wrote an important book* is compatible with the proposition *The traffic
problem in New York is insoluble*, and both of these with the
proposition *Some schoolchildren like to study logic*. But this
bare compatibility is uninteresting, because nothing else
follows from it. That is why the disconnected remarks of
a Mrs. Nickleby or a Miss Bates are apt to be boring, and
why some old gentlemen's stories are pointless. The
relation of bare compatibility cannot afford a basis for
inference, nor provide the material for a joke. Nevertheless,
it is necessary to notice that such a relation does hold
between certain propositions. Two propositions thus related
are said by logicians to be *independent*.

Compatibility is a very general relationship between propositions. The logician is mainly interested in the more
specific relationships which may hold between propositions
which are compatible. *Independence* is one such relationship, but there are several others which must be

[1] *Shorter Oxford English Dictionary.*

distinguished. Another of these is *equivalence*. Examples are: *All poets are sensitive to criticism, No poets are insensitive to criticism; If Roosevelt abolishes war debts, the Americans will be displeased, Either the Americans will be displeased or Roosevelt will not abolish war debts.*[1] An examination of these two examples will show that, in each case, the truth of the second proposition follows from the truth of the first, and conversely; the falsity of the second proposition follows from the falsity of the first, and conversely. Hence, to assert either entails the assertion of the other. Thus two propositions are *equivalent* if one entails the other, and conversely.

Two propositions are also compatible if one can be inferred from the other, even though the other cannot be inferred from the former. Thus, *Some poets are sensitive to criticism* can be inferred from *All poets are sensitive to criticism*. The reverse inference is not, however, permissible. Hence, the relation of the first of these two propositions to the second is different from the relation of the second to the first. The two relations must, then, be distinguished. One proposition is *sub-implicant* to a second if the first can be inferred from the second, but not conversely. In this case, the second is *super-implicant* to the first. The relations are called respectively *sub-implication* and *super-implication*.

Two propositions may be compatible although neither can be inferred from the other, and yet they are not independent. This is so when the two propositions are so related that they cannot both be false and may both be true. For example, *some stupid people are obstinate* and *some stupid people are not obstinate*. If we know that one of these propositions is false, we can infer that the other is true; but if we know that one is true, we cannot infer that the other is true, nor that it is false. Both possibilities remain open. Fortunately this is plain common sense. Logicians have unfortunately invented the inappropriate word '*sub-contrary*' to express this relation.

If two propositions are incompatible, the truth of one

[1] Examples of equivalent propositions will be found in pp. 35–6 below.

cannot be inferred from the truth of the other. Since, however, of two incompatible propositions one at least must be false, knowledge that one of them is true enables us to infer that the other is false. We must distinguish between incompatible propositions which are *contrary* and those which are *contradictory*. Two incompatible propositions are *contrary* if neither need be true and both cannot be true. Examples are: *No economic theories are sound, All economic theories are sound; Tobias Fortescue always tells lies, Tobias Fortescue never tells lies*. To each of these pairs of propositions there is obviously an alternative. Possibly Tobias Fortescue sometimes tells lies and sometimes speaks the truth. Two incompatible propositions are *contradictory* if one must be true and one must be false. It follows that from knowledge of the truth, or falsity, of the one the falsity, or truth, of its contradictory can be inferred. Examples are: *Whoever trusts a politician's promises shows himself to be foolish, One may trust a politician's promises without showing oneself to be foolish; St. Paul's Cathedral is smaller than St. Peter's at Rome, St. Paul's is either the same size as, or is larger than, St. Peter's at Rome.* A proposition is denied when either its contradictory or a contrary to it is asserted.

The seven relations, with regard to inferability, which may hold between two propositions are, then: (1) independence, (2) equivalence, (3) sub-implication, (4) super-implication, (5) sub-contrariety, (6) contrariety, (7) contradiction. When (1) holds, no inference is possible; when (2), (3), or (4) holds, at least one of the two propositions implies the other; when (5), (6), or (7) holds, neither proposition implies the other, but—under the various conditions specified above—it is possible to infer something with regard to the truth or falsity of one proposition given knowledge of the truth or falsity of the other proposition.

We have next to consider affirmation and denial. These notions are familiar to every one. A question can usually be so phrased that it admits of being answered by a 'Yes' or by a 'No', although sometimes we may be in doubt as to which answer is correct.[1] The *Yes*-answer is in effect an affirmation;

[1] See pp. 76 seq.

the *No*-answer is in effect a denial. For example, the reader may be asked, 'Is it worth while to study logic?' If he answers, 'No', then he is in effect saying that it is *not* worth while to study logic; if he answers, 'Yes', then he is in effect saying that it *is* worth while to study logic. It is what the *question* is about that determines whether the answer is an affirmation or a denial. The question, cited as an example, is about *the worth-whileness of studying logic.* The questioner wants to know whether the property of *being worth while* belongs, or not, to the study of logic. A denial that the property belongs might be expressed by the sentence, 'It is a waste of time to study logic', or by the sentence, 'Studying logic is unworthwhile'. These *sentences* are affirmative, but the speaker who uses one of them to express his answer to the given question is making a denial. The distinction between affirmative and negative sentences derives its significance from the distinction between affirmation and denial. In affirming, or denying, we *use* sentences, but what we affirm, or deny, is not a sentence but what the sentence is used to express. In answering a question the sentence is used to express a proposition. A proposition expressed by an affirmative sentence is usually called an *affirmative proposition*; one expressed by a negative sentence is called a *negative proposition*. In dealing with propositions out of the context in which they may be asserted, this is a convenient procedure. But it must not mislead us into supposing that the *same* state of affairs cannot be referred to both by a negative and by an affirmative proposition. On the contrary, every affirmative statement can be translated into a corresponding negative statement (and conversely), which is equivalent to the original. For example: *Question*—'Are philosophers consistent?' *Answer*—'No. Some philosophers are inconsistent.' The answer might have taken the form, 'Some philosophers are not consistent.' In the first form, *being inconsistent* is affirmed of *some philosophers*; in the second form, *being consistent* is denied of *some philosophers*. Both statements refer to the same characteristic of these philosophers. The negative statement must deny the possession of the *opposite* property to that which the affirmative statement asserts to be possessed. Such an opposite property is called a *contradictory* property, e.g. *consistent* and *inconsistent* are contradictory properties. But *not consistent* is equivalent to *inconsistent*. *Which* of the two contradictory properties we shall assert of a subject depends upon the facts of the case (or rather, what we believe to be the facts); but whether we make our assertion in an affirmative or a negative sentence will depend upon the form of the question we are answering (e.g. 'Are philosophers consistent?' or 'Are philosophers inconsistent?')

It follows, then, that we have not denied the fundamental distinction between affirmation and denial. On the contrary, we have insisted upon it in maintaining that to *affirm* any property of something is equivalent to *denying* the posssession of the *contradictory* property. Two propositions thus related are said to be *obverses* of each other; the process of drawing one of these propositions from its equivalent is called *obversion*. An inference from a single proposition to another implied by it is called an immediate inference. The name is not fortunate; it may, however, serve to remind us that in asserting any proposition whatever we are committed to *other* assertions, namely, to whatever is implied by the original proposition whether we happen to have noticed these implications or not.

Statements about things of a certain sort are, we have seen, statements about *classes*, e.g. *philosophers*. With regard to any property which we could significantly think of as belonging, or not belonging, to philosophers, there are three possibilities. *Every* philosopher might possess the property; or *none* might; or *some* might possess it and *some not*. For example, Some philosophers are hot-tempered and some are not; No philosophers are consistent; All philosophers are liable to headaches. In *denying* that *every* philosopher possesses a given property, we commit ourselves to the assertion that *some* do *not* possess it. In short, to deny a given statement is to affirm its contradictory.

In the preceding discussion we have been making use of two fundamental logical principles, which together determine the nature of contradiction. These principles have received technical names. They may be stated as follows:

I. Principle of Non-Contradiction: *Given any proposition, P, then P cannot be both true and false.*
II. Principle of Excluded Middle: *Given any proposition, P, then P is either true or false.* A third alternative is excluded, since there is no mean between truth and falsity.

With regard to a given proposition we may not *know* whether it is true or whether it is false; we know, however, that it must be one or the other and that it cannot be both.

These principles may be stated in a form which is directly applicable to the possession by an object of a property. Let O be any object, and F any property which could be significantly asserted of O.[1] It will be convenient to represent the property contradictory to F by 'non-F'. Then we get:

I. O cannot possess both F and non-F.
II. O must possess either F or non-F.

The class *philosophers* was taken *as an example*, just as in the preceding chapter we took *psychoanalysts, seals, mammals, blasting operations, ships.* We could have taken *any other class.* The characteristics distinguishing one sort of thing from another sort have throughout been irrelevant to our discussion, since we have been concerned only with the *form* of the assertion. What we have said applies quite generally to *any* class and to *any* property. We can show that our statement is quite general by using a *symbol* to stand for *any class,* just as in algebra we use a symbol to stand for *any number.* It was in this way that we used symbols in formulating the two principles of non-contradiction and excluded middle.

[1] It is not significant, i.e. does not make sense, to assert some properties of some objects. For example, it is nonsensical to say, 'Some courageous acts are triangular', and equally nonsensical to say 'Some courageous acts are non-triangular'. The property of *being triangular* could be significantly affirmed, or denied, only of that which has shape.

Using X and Y to stand for any two different classes, we can use *All X's are Y's* to represent *All philosophers are thinkers*, or *All civil servants are patriotic*, or *any other* statement of *the same form*. In fact, what *All X's are Y's* represents just is the *form*, which is common to ever so many different statements, namely, to those which assert that every member of one class is included in some other class. It is this sameness of form which alone is relevant to the purposes of a logician. Similarly we can represent by *Some X's are Y's* what is common to all statements, to the effect that some members of a given class are included in some other class.

A proposition of the form *All X's are Y's* is called a *universal* proposition; one of the form *Some X's are Y's* is called a *particular* proposition. The distinction between universal and particular propositions is very important. Upon it depends the fact that an inductive generalization from *some* members of a class to *all* is an inference going beyond the evidence. It is not, however, the case that in asserting that some X's are Y's we always intend to assert that *only* some are. If it were, then, we could *never* pass from the assertion that *Some psychoanalysts lack humour* to the assertion that *all do*. Although the evidence may be sufficient only to justify the assertion about *some*, nevertheless the universal assertion might be in fact true. Thus the assertion of *Some X's are Y's* is compatible with the assertion of *All X's are Y's*. The force of 'some' is selective; its use enables us to make a partial generalization, leaving open the possibility that a universal generalization could also be truly asserted. In using symbolic expressions it is important that each symbol should have a fixed reference. Accordingly, we have to decide whether to interpret 'some' in the sense 'some, it may be all', or in the sense 'some, but not all'. For the reasons given, it is convenient to adopt the former interpretation. Then, if we wish to say that some but not all X's are Y's, we assert the compound proposition. *Some X's are Y's and some X's are not Y's.* It is, indeed, easy to see that *Only some socialists are Marxians* denies *both* that all socialists are Marxians *and also* that no socialists are

Marxians. The sentence is grammatically simple, but the proposition expressed is compound.

These various statements about classes can be symbolically represented as follows:

(1) *All*   *X's are Y's.*
(2) *No*   *X's are Y's.*
(3) *Some* *X's are Y's.*
(4) *Some X's are not Y's.*

We may briefly summarize what has already been said about these four forms. (1) and (2) are forms of universal or unrestricted, generalizations. (3) and (4) are forms of partial, or restricted, generalizations. (1) and (3) are affirmative; (2) and (4) are negative. Negative propositions of these forms can be regarded as denying inclusion, i.e. as asserting exclusion of one class from another. Looked at from this point of view, we can see that the four forms are derived from the fact that, with regard to any class X, we can assert that X is either wholly or partially included in, or wholly or partially excluded from, the class Y.

Any one of these four propositions can be regarded as consisting of *two terms* combined in the way appropriate to its special form. For example, *No civil servants are members of Parliament* may be said to be about *civil servants* and *members of Parliament*. Since what is said about them is that the one class is wholly excluded from the other, it is clear that *No members of Parliament are civil servants* is equivalent to the original proposition. In the context of a discussion, which of the two propositions we chose to assert would depend upon the question determining the direction of our thinking. If we were asking: 'Are any civil servants members of Parliament?' we should probably assert the original proposition: if we were asking 'Are any members of Parliament civil servants?', we should assert the second. The term which comes first is often called the *subject*, the second term the *predicate*. In this usage of the words 'subject' and 'predicate' we think of the proposition as asserting something about something. *That which is asserted* about something is the *predicate*; *that about which* something is asserted is the *subject*.[1]

It should be observed that we have used 'about' in two different senses. In the former sense the proposition is 'about' *both the terms*; in the latter it is 'about' *the subject-term*. The narrower sense is due to the fact that the question which determines our thinking is a question concerning the subject-term. In the context of a discussion the subject-term is not always stated first; *which* of the terms is subject must be decided by reference to a question which the given proposition might be regarded as answering. In considering propositions, taken in isolation for the purposes of example, we shall assume that the subject-term is the first term.

We have seen that in the case of the proposition *No civil servants are members of Parliament*, the subject- and predicate-terms can be interchanged without alteration in the truth, or falsity, of the proposition asserted. In general, any proposition of the form *No X's are Y's* is equivalent to *No Y's are X's*. These propositions are said to be simple converses of each other. Their equivalence is due to the fact that *complete exclusion* is a reversible—or, as logicians say, a symmetrical—relation, so that in wholly excluding one class from a second, the second is *ipso facto* wholly excluded from the first. Notice that *every* member of the class indicated by the subject-term is asserted to be excluded from the class indicated by the predicate-term. Where the reference is to *every* member of the class, the term is said to be *distributed*.

A proposition of the form *All X's are Y's* asserts the relationship of *complete inclusion*. This relation is not symmetrical and we cannot therefore apply the method of simple conversion, i.e. conversion by interchanging X and Y without making any other change. *All seals are mammals* is not equivalent to *All mammals are seals*. The former proposition implies that seals are *some* of (but not that they are *all*) the creatures classified as mammals. Notice that in *All X's are Y's*, whereas the subject is distributed, the predicate is not. The particular affirmative form *Some X's are Y's* involves the relationship of *partial inclusion*, and this

[1] Here 'predicate' is used in a wide sense.

relation *is* symmetrical. The prefix *some* restricts the reference of the subject-term, just as the reference of the predicate is restricted. *Some X's are Y's* is thus equivalent to *Some Y's are X's*.

The particular negative form *Some X's are not Y's* involves the relationship of *partial exclusion*. Such a proposition cannot be simply converted. For example, *Some fish-eaters are not cats* is compatible with *Some cats are not fish-eaters*, but it is also compatible with *All cats are fish-eaters*. Since the original proposition leaves open both these possibilities, to infer the simple converse would be to go beyond the evidence. Notice that in *Some X's are not Y's* the predicate, unlike the subject, is distributed, for such a proposition might be expressed by saying 'Part of class X is excluded from the *whole* of class Y.'

It should be clear that whether a proposition is simply convertible, or not, depends upon the distribution of the terms. If *both* terms are distributed, or if *neither* term is distributed, then the proposition is simply convertible. But if one term is distributed, whilst the other is not, then the proposition has no simple converse. In this case, however, the proposition has a negated converse equivalent. For example, *All polite people are tactful* is equivalent to *No untactful people are polite*. In totally including those who are polite in the class of those who are tactful, we *ipso facto* exclude, the *untactful* from *the polite*. Likewise, *Some great statesmen are not free from vanity* is equivalent to *Some who are not free from vanity are great statesmen*. If the reader looks back to what was said about obversion, he will see that these equivalents are converted obverts of each other.[1]

The converse equivalents may be summed up in the following schema:[2]

[1] A converted obvert is sometimes called a *contrapositive*. Since *every* proposition can be obverted, the contrapositive can be again obverted. The obverted contrapositive of *All polite people are tactful* is *All untactful people are impolite*. The reader will easily be able to derive other examples. For a fuller discussion, see L. S. Stebbing: *A Modern Introduction to Logic*, Chap. V, § 2.

[2] In this schema, the sign ≡ is used as a shorthand symbol for 'is equivalent to'. It will often be found convenient to use this abbreviation.

Simple Converse Equivalents.

No X's are Y's   ≡ No Y's are X's.

Some X's are Y's   ≡ Some Y's are X's.

Negated Converse Equivalents.

All X's are Y's   ≡ No non-Y's are X's.

Some X's are not Y's   ≡ Some non-Y's are X's.

The reader should observe that the equivalence of the propositions, in each of these pairs, results from the distributive reference of their respective terms. In deductive inferences distribution is of fundamental importance, since to infer a proposition, containing a distributed term, from a premiss in which that term was given undistributed would be to go beyond the evidence. The facts concerning distribution, in the case of each of the four forms of propositions given above, may be summarized as follows: (1) The *predicate-term* is distributed in a negative, but undistributed in an affirmative proposition; (2) the *subject-term* is distributed in a universal, but undistributed in a particular proposition.

Bearing these points in mind, we may ask what exactly is the information given us by the statement that *All polite people are tactful.* It informs us that:

(1) If anyone is polite, he is tactful.[1]

(2) If anyone is not tactful, he is not polite.

It leaves open the two possibilities:

(i) That someone is tactful although not polite.

(ii) That no one is tactful without being polite.[2]

---

[1] This is just another way of saying 'All polite people are tactful', whilst (2) is another way of saying 'All untactful people are impolite.' Hence, (1) and (2) are contrapositives of each other, and are thus equivalent.

[2] The two possibilities are left open because the predicate-term of the original statement is not distributed. Accordingly, whilst *the polite* are restricted to *the tactful*, the converse is not the case. The reader will find it worth while to convince himself of the truth of these contentions, and to see that their truth is a consequence of what was said in the discussion of obversion and the converse equivalents.

Not both these possibilities can be realized, but the given statement about *polite people* does not tell us which is in fact the case. It is clear, however, that, if we know with regard to a certain person, say Ramsay MacDonald, that he is polite, we can deduce that he is tactful. Again, should we happen to know with regard to another person, say Tobias Fortescue, that he is not tactful, then we could deduce that he is not polite. But if we knew only that he is not polite, we could not (on the sole basis of the generalization about polite people) deduce that he is tactful, nor that he is not. That the latter deduction would be invalid follows from the fact that possibility (i) was not *excluded* by the original statement.

The arguments suggested above are examples of a very common form of argument, which has been technically called *syllogism*. A syllogism may be regarded as essentially consisting in the application of a generalization (or a general rule) to a specified case in order to deduce a result. We may begin by considering an example:

(Rule)   All aviators are intrepid.
(Case)   Amy is an aviator.
(Result) ∴ Amy is intrepid.

Suppose we deny that Amy is intrepid. Then we must, in consistency, deny *either* that she is an aviator *or* that all aviators are intrepid. Then we get:

(Denial of Result)   Amy is not intrepid.
(Case)   Amy is an aviator.
(Denial of Rule) ∴ Some aviators are not intrepid; provided that we keep to the assertion of the Case. If, however, we are prepared to maintain that all aviators are intrepid, we must accept the following argument:

(Rule)   All aviators are intrepid.
(Denial of Result)   Amy is not intrepid.
(Denial of Case) ∴ Amy is not an aviator.

The reader will have no difficulty in seeing that each of these three arguments is valid, i.e. the acceptance of the
4

premisses entails the acceptance of the conclusion. Accordingly, the denial of the conclusion entails the denial of *at least one* of the premisses. Thus, denial of the Result, combined, with acceptance of the specified Case, entails denial of the Rule; denial of the Result, combined with acceptance of the Rule, entails denial of the specified Case. This point may be put in a different, but equivalent way. The three propositions: (1) *All aviators are intrepid*, (2) *Amy is an aviator*, (3) *Amy is not intrepid*, cannot be true together. The combination of *any two* of them entails a conclusion which contradicts the omitted proposition.

The principle in accordance with which the first of these three arguments proceeds may be formulated as follows: *Whatever can be asserted (affirmatively or negatively) of any member of a given class can be likewise asserted of any specified member.* This is called the *Applicative Principle*, since it permits us to apply to a specified case whatever is asserted of *every* case in general.[1] The Principle yields the symbolic form:

> If   Anything which is a member of X has F (or not),
> and A is a member of X;
> then A has F (or not).[2]

The bracketed 'or not' shows that the form is valid whether the property be affirmed or denied of the members of X, but that it must be *in like manner* affirmed or denied of A.

An allied principle—which may be called the *Principle of Excluding an Individual from a Class*—covers the case of deducing that Tobias Fortescue is not polite since he is not tactful, and all polite people are tactful. The Principle may be formulated as follows: *If a given individual lacks (or possesses) a property which any member of a certain class possesses (or lacks), then that individual is not a member of that class.* This yields the form:

---

[1] This Principle is also called the *Principle of Substitution*, for it is the Principle in accordance with which *values* can be substituted for variables in, for example, $'(a+b)(a-b) = (a^2 - b^2)'$.

[2] Here (and subsequently) X stands for any class, F for any property, A for any *specified* individual.

If   Anything which is a member of X has F (or not),
and A has not F (or has);
then A is not a member of X.

We may now consider another type of syllogism, in
which the specified case may be regarded as replaced by a
set of cases of the same sort, this set being a class falling
within a wider class. For example:

All intrepid people are admirable.
All aviators are intrepid.
∴All aviators are admirable.

No civil servants are M.P.s.
Some who direct the Government are civil servants.
∴Some who direct the Government are not M.P.s.

The principle in accordance with which this reasoning
proceeds may be formulated as follows: *Whatever can be
asserted of every member of a class can in like manner be
asserted of every sub-class contained in that class.* This
principle has been named the *Dictum de omni et nullo*. It
yields the form:

If   Every Y is Z (or not),
and Every (or some) X is Y;
then Every (or some) X is Z (or not).

An examination of this form shows that a connexion is
established between X and Z on the ground of the connexion
between Y and Z, on the one hand, and between X and Y,
on the other.   Accordingly, Y, which occurs in both
premisses but not in the conclusion, is called the *middle
term*; X and Z are called the extreme terms.   Unless the
middle term is distributed in at least one of the premisses,
no connexion is secured; at least one of the extreme terms
must be given as having a relation to the whole of class Y.
An example may make this clear:

Some intelligent people are witty.
All civil servants are intelligent.
∴All civil servants are witty.

In this argument the conclusion does not follow, and indeed the premisses do not even warrant the weaker conclusion 'Some civil servants are witty.' The intelligent witty people may not include any of the civil servants; it might even be the case (so far as the evidence provided by the premisses goes) that the civil service, whilst requiring intelligence, deadens wit. The conclusion is *consistent* with the premisses, but not *entailed* by them. This may be exhibited by a diagram in which intelligent people are collected into one circle, witty people into another. The first premiss ensures that these circles must at least overlap. Thus we get:

The state of affairs asserted by the second premiss is consistent with the three cases: (i) *all* civil servants are in the overlapping portion; (ii) *none* are; (iii) *some* are and *some* are *not*. Thus these premisses do not suffice to tell us which of the three cases is correct. If, however, intelligence were asserted to be a *sufficient* condition of wit, the first premiss would become *All intelligent people are witty*, and the arrangement of the circles would become

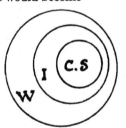

With the emended premiss the syllogism is of the form,

> Every Y is Z.
> Every X is Y.
> ∴ Every X is Z.[1]

[1] In some cases the circles might coincide.

This syllogism is well known to logicians under the proper name *Barbara*. In order to establish a universal affirmative conclusion both premisses must be universal affirmatives, and the terms must be arranged as in the above schema. If the first premiss is negative, the conclusion must also be negative. The second premiss, being the statement that a certain sub-class is contained in a wider class, is affirmative. If this premiss makes an assertion about *some* members of this sub-class, X, then the conclusion must also be particular.

The arrangement of the terms in the form yielded by the *Dictum de omni et nullo* is the commonest arrangement. It will be observed that the middle term is predicate in the second premiss, and subject in the first premiss. Valid syllogisms need not conform to this arrangement. For example:

No good citizens are selfish.
All people who leave litter are selfish.
∴ All people who leave litter are not good citizens.

If for the first premiss we were to substitute its converse equivalent, this syllogism would be in the form appropriate to the *Dictum de omni et nullo*. By using the various equivalents, such rearrangement of the position of the terms is usually possible, but it is not necessary. Other *Dicta*, directly applying to other arrangements, can be formulated.[1] We shall not state these *Dicta* here. The reader will be able to formulate a *Dictum* applying to the case where the middle term is predicate in both premisses, if he considers the Principle of Excluding an Individual from a Class, which was given above. It is sufficient here to state briefly the Rules which guarantee the validity of any syllogism. These are: (1) the middle term must be distributed in at least one of the premisses; (2) if a term is distributed in the conclusion, it must be distributed in the corresponding premiss; (3) at least one premiss must be affirmative; (4) if one premiss is negative, the conclusion must be negative; (5) if both

---

[1] For the statement of these *Dicta*, and a full discussion of other arrangements of terms, see L. S. Stebbing: *A Modern Introduction to Logic*, Chap. VI.

premisses are affirmative, the conclusion must be affirmative. These Rules will apply both to class syllogisms, and also to those syllogisms in which one term is an individual, notwithstanding the difference in the logical properties of the two types of syllogism.[1]

The validity of the class syllogism depends upon the nature of the relation of *inclusion*. If a plane figure (e.g. a circle, or square, or triangle, etc.) is wholly included in a second which is wholly included in a third, then the first is wholly included in the third. A relation having such a property is called *transitive*. This property of *transitiveness* is extremely important, for upon it depends the validity of all deductive arguments, except those which depend upon the Applicative Principle. The property may be described as follows: A relation is *transitive* if it is such that, given that the relation relates X to Y, and also relates Y to Z, then it follows that X must be related to Z by this relation. Familiar examples of transitive relations, in addition to *inclusion in*, are *equals, greater than, exactly contemporary with, older than, more pious than*. If a set of policemen are arranged in order of *height*, beginning with the shortest and ending with the tallest, then (on the assumption that no two were exactly the same height) we could pass down the row, picking out any two of which it would be true to say that the one nearer the end was taller than the one nearer the beginning of the row. This property of transitivity yields a principle—called by William James, 'the axiom of skipped intermediaries'. He says, 'Symbolically we might write it as $a < b < c < d$ ... and say, that any number of intermediaries may be expunged without obliging us to alter anything in what remains written.'[2]

Implies is a transitive relation. It is in virtue of this property of implication that it is possible to combine a set of propositions, of which the first implies the second, the second

---

[1] It is not possible to deal properly with these differences here: The reader will find a discussion of the point in L. S. Stebbing. op. cit., p. 97.

[2] *Principles of Psychology*, vol. II, p. 646. Cf. L. S. Stebbing: op. cit., p. 173.

implies the third, and so on, into a system such as the Euclidean system of geometry. Many of our ordinary arguments are intended to consist of a set of propositions which successively imply others. Usually we skip the intermediate steps, and may sometimes be led to asserting that a proposition implies another, although, had we made the steps explicit, we should have seen that the chain of implications had been broken.

We have so far considered deductive arguments in which (i) one term was an individual;[1] (ii) the three terms were all class terms. We have now to consider arguments in which all the terms are individuals. A proposition of which an individual is the subject is a *singular* proposition, e.g. *Hitler is aggressive, Hitler admires Mussolini*. The first of these is a subject-predicate proposition; the second is known as a *relational proposition*. A relational proposition is one in which two, or more, terms are stated to be related by some definite relation. A relational argument consists of relational propositions, e.g.:

(i)   Galileo preceded Newton.
      Newton preceded Einstein.
      ∴ Galileo preceded Einstein.

(ii) My sister runs as fast as Tom.
      Tom runs as fast as your brother.
      ∴ My sister runs as fast as your brother.

(iii) A is father of B.
      B is father of C.
      A is father of C.

The reader will observe that (i) and (ii) are valid, whilst in (iii) the conclusion does not follow. The premisses here imply the conclusion *A is grandfather of C*. The relation *father of* is not transitive; on the contrary, it is *intransitive*, i.e. it is such that if it holds between A and B, and between B and

---

[1] In this book 'an individual' means exactly what the reader will take it to mean. In this sense a *class* is clearly *not* an individual, but a collection of individuals, or a collection of sub-collections.

C, then it cannot hold between A and C. Similarly, whilst *older than* is transitive, *older by one year than* is intransitive. From *A is older by one year than B* and *B is older by one year than C*, there follows the conclusion *A is older by two years than C*. We see this as soon as we understand what 'older than' and 'older by one year than' *mean*. A knowledge of the system of relations which render such arguments valid constitutes part of our common stock of information. The plain man is not likely to make serious mistakes simply because of a failure to know whether a relation is, or is not, transitive. If he does, logic cannot help him. Are the friends of our friends also our friends? It is not for the logician to decide whether *friendship* is a transitive relation.[1]

There remains a set of deductive forms, of common occurrence in ordinary discourse, in which one of the premisses is a compound proposition. A compound proposition is a combination of two, or more, propositions each of which is separately assertable. A simple proposition is one which is not compound.[2] It will be sufficient for our purpose to consider four modes of combining simple propositions. The simplest combination is that effected by the logical conjunction *and*. The proposition *Drake played bowls and subsequently he fought in the Armada* consists of two simple propositions; it is true if both these propositions are true; false, if either of them is false. The constituent propositions are logically independent; the conjunctive proposition merely asserts that both are true. The other three modes of combination are effected by the three combining forms, *If . . . , then . . .; Either . . . or. . .; Not both. . . .*

A proposition such as *If Hitler defies Europe, the League ought to intervene*, is called *Hypothetical*. It asserts that a certain condition (*Hitler defies Europe*) has a certain consequence (*the League ought to intervene*). The condition is called the *Antecedent*; the consequence is called the

[1] The reader interested in the logical treatment of relations should consult L. S. Stebbing: op. cit., Chap. X.
[2] We have treated propositions of the form *All X is Y*, etc., as *simple*. There are objections to this procedure, but the proper treatment of this topic would take us too far afield. See L. S. Stebbing: op. cit., Chap. IV.

Consequent. The reader should observe that the hypothetical proposition may be true although *neither* of its constituents is true; it is false if the antecedent is true although the consequent is false.

A proposition such as *Either you are an optimist or I am misinformed as to the facts* is called *Alternative*. This name is self-explanatory. An alternative proposition is true if at least one of the alternatives is the case; it is false if neither is the case. Notice that 'either . . . or . . .' does not exclude the possibility that both of the alternatives are true. *He is either clever or hard-working* leaves open the possibility that he is both; it only excludes the possibility that he is neither. There are, of course, occasions when we use 'either . . . or . . .' in an exclusive sense. For example, the purpose of a waiter who said, during war-time, 'You can either have cheese or a meat-course' was to inform us that we could not have both. But in what follows we shall use 'either . . . or . . .' in the non-exclusive sense, for the exclusive sense can be expressed unambiguously by 'not both . . .'

A proposition such as *It is not the case both that a gale is blowing and that it is safe to swim in the bay*, is false, provided that each of the disjoined propositions is true, for this compound proposition asserts that at least one of its constituents is false. It is called a *disjunctive* proposition.

The conjunctive form does not give rise to any inferences which are not redundant. Each of the other three forms can afford a premiss in an inference, as the reader may have observed from the statement of its significance. This may best be shown by means of an example:

'If Roosevelt's Recovery Campaign succeeds, then the economic system of America will be a form of "controlled capitalism". In that case it follows both that industry will be co-ordinated and unified and also that Labour will be in a new relation to employers. If it succeeds, then either Great Britain must follow the example of America, or else she will become bankrupt. The latter alternative is unthinkable. Hence we may conclude that Great Britain would have to adopt the American plan.'

The reader should have no difficulty in assigning each of the single propositions in the above argument to its appropriate form. The argument is not fully stated. The passage is argumentative because certain of the propositions are said (explicitly or implicitly) to imply others. But the implied propositions are not formally set out. Hence, in the strict sense this passage does not present *a formal argument*; it presents *the material for one*. This is the mode of procedure adopted in ordinary discourse and in most argumentative discussion. It is not formal reasoning, but it has, for the instructed hearer, the force of formal reasoning. Whenever we use a 'therefore', 'hence', 'consequently', 'we conclude that', 'it follows from', 'since', or 'because', we are stating an argument, the validity of which depends upon its being an example of a *form of implication*. We seldom state fully the premisses upon which the validity of our argument depends. To do so would be tedious and is often unnecessary. The omission, however, sometimes leads us into drawing erroneous conclusions, as we shall see later.

To return to the example. The first statement does not assert that Roosevelt's Campaign *will* succeed; it asserts a consequence of its success, *should it succeed*. The second statement asserts a conjunction of two other consequences of its success. The third statement asserts a further consequence, in the form of an alternation; it does not assert *which* alternative would be realized, but merely that *at least one* will be. The next statement cuts out one alternative. The last statement explicitly *draws the conclusion* which follows from the denial of one alternative.

The formal rules of such arguments as the above are easy to understand. We shall, accordingly, state these rules briefly, and shall adopt the convenient device of using P to stand for *any one* proposition, Q for *any other* proposition.

Hypothetical argument. This has two forms:

(I) If P, then Q	(II) If P, then Q.
P	not Q
∴ Q	∴ not P.

Rule (1): Affirm the Antecedent; the affirmation of the Consequent follows.

Rule (2): Deny the Consequent; the denial of the Antecedent follows.

It should be observed that nothing follows from the denial of the Antecedent, nor from the affirmation of the Consequent. This follows from the fact that the Antecedent and the Consequent cannot be simply interchanged. *If it keeps fine, he will go out*, is not equivalent to *If he goes out, it will keep fine*. He may be forced to go out, even though he dislikes wet weather. Had the original proposition been *Unless it keeps fine, he will not go out*, then the denial of the condition *it keeps fine* would entail that he does not go out. But in this case the antecedent is *If it does not keep fine*, for 'Unless' means 'If . . . not'. Hence, the denial of fine weather constitutes an affirmation of this Antecedent. The fallacy of *Affirming the Consequent* (i.e. asserting the Consequent to be true and, thence concluding that the Antecedent is true) is very common. This fallacy involves assuming that the condition specified in the Antecedent is *the only* condition whose fulfilment would ensure the truth of the Consequent. But this is to go beyond what is asserted by a hypothetical proposition.

(III) *Alternative Argument.*
> Either P or Q
> not P.
> ∴Q.

Rule: Deny one alternative; the affirmation of the other alternative follows.

Since the alternatives are not exclusive, nothing follows from the affirmation of one alternative.

(IV) *Disjunctive Argument.*
> Not both P and Q.
> P.
> ∴ not Q.

Rule: Affirm one of the disjoined propositions; the denial of the other follows.

The denial of one proposition in the disjunction does not warrant the affirmation or denial of the other. Both may fail to be the case, since the premiss merely asserts that *at least one* is not the case.

There are various complicated arguments consisting of combinations of different compound propositions. These, however, do not exhibit any other logical principles than those with which we have dealt. We shall notice one form only, the Dilemma, which is familiar to educated people. Its use is primarily rhetorical, since it affords an effective argumentative device. A dilemma is a form of argument the purpose of which is to prove that an unwelcome conclusion follows from either of two alternatives. It contains two premisses. The first consists of two hypothetical propositions *conjunctively affirmed*; the second is an alternative proposition whose constituents are either the antecedents of the hypothetical propositions or the contradictories of their consequents. One example will suffice:

' "If you succeed in your Recovery Campaign", a well-known American is said to have remarked to Mr. Roosevelt, "you will be known as America's greatest President. If you fail you will be known as the worst." "No," Mr. Roosevelt is reported to have answered; "if I fail, I shall be known as the last President." '

This, again, is not an argument; it presents the material for an argument. We may formulate the argument. abbreviating the statements:

> If Roosevelt succeeds, he will be known as America's greatest President, and if he fails, he will be known as America's last President;
> But, either he will succeed, or he will fail;
> Therefore, either he will be known as America's greatest President, or as its last.

The form may be symbolized, using simple capital letters to stand for simple propositions, as follows:

> If P, then Q, and if not P, then R;
> But either P or not P;
> ∴ either Q or R.

In this form the Antecedents of the two hypothetical propositions are contradictories; hence, one of them must be the case. This is not always so. We must recognize four more forms:

(i)   If P, then Q, and if R, then T;
        But either P or R
        ∴ either Q or T.

(ii) Another form is yielded by the alternative denial of the Consequents, entailing the alternative denial of the Antecedents:

> If P, then Q, and if R, then T;
> But either not Q or not T
> ∴ not P or not R.

(iii)   If P, then Q, and if R, then Q;
         But either P or R
         ∴ Q.

Here the Consequents are common, and are asserted to follow from both of two Antecedents.

(iv)   If P, then Q, and if P, then R;
         But either not Q or not R
         ∴ not P.

A dilemma is inconclusive if the alternatives given in the second premiss do not exhaust all the possibilities. For example:

> If my football team wins, I'll lose my bet,
> and if it loses, I'll lose prestige;
> But my team must win or lose
> ∴ I must lose either money or prestige.

This argument is inconclusive because it ignores the fact that a football match may result in a draw.

The validity of our deductive reasoning depends upon

formal principles so easily apprehended that the plain man may think them too obvious to need statement; such as the Principles considered in this chapter, viz. Non-Contradiction, Excluded Middle, Applicative Principle, *Dictum de omni et nullo*. No one who understands these Principles is likely to deny them to be true. But they are not the less important because they are obvious. Unless the conclusiveness of our reasonings could be shown to depend upon such obviously true principles, we should have no good reason for holding that any of our inferences were *conclusive*. As logicians we want to know what are the principles which guarantee the validity of our reasoning. These principles must be formal in virtue of the fact that conclusive reasons in one case must be conclusive reasons in any other case of the same form. If, therefore, the conclusion of valid reasoning is false, its falsity must be sought in the material conditions; at least one of the premisses must be false. Valid deductive argument consists in drawing a conclusion from a premiss, or a set of premisses, which together entail the conclusion. If the premisses are true, the conclusion must also be true.

CHAPTER IV

AMBIGUITY, INDEFINITENESS, AND RELEVANCE

'The line of all progress in disputes is towards definiteness—definiteness of issue, definiteness in the conception
of the facts appealed to, and of the precise meaning of
those facts.'
                                                    A. SIDGWICK

IN discussing examples of reasoning in the last chapter
we did not pause to inquire whether we clearly understood the sentences used to express our statements. When
we converse or argue with other people, when we read or
write—in short, on nearly every occasion of reflective
thinking—we use language. A language consists of a set
of symbols capable of being combined in various ways in
order to express beliefs about different states of affairs.
Symbols are signs used by some one, in accordance with a
convention, to refer to something. Words are one kind of
symbol. They are *sounds* (in spoken language) or *marks*
(in written language) *used* by those who speak the language.
Words have *meaning*, but the sound, or mark, has not meaning of itself; it becomes a *sign*, and thus aquires significance
or meaning through its use in accordance with a convention.
It is important to stress this conventional element since it is
apt to be forgotten. We then come to think of the *mark*, or the
sound, as the *word*, and suppose that we can precisely determine the meaning of the word by looking up the mark in the
dictionary. Thereupon, we fall into the mistake of supposing that words can be exhaustively divided into those which
are ambiguous and those which are free from ambiguity.
Both these suppositions are mistaken. Meaning belongs
only to *the sign as used*. Hence, to know what a given word
means, we must know *how* it is being used in the context
in which the speaker[1] is using it. Since the contexts in

---

[1] Wherever we say 'a speaker' we could add 'or writer'; likewise,
'hearer', or 'reader'. For brevity, we shall henceforth speak of 'the
speaker'. This is in conformity with the ordinary usage of 'speak'
in the last sentence.

which some words are used present considerable similarity, these words have a comparatively fixed reference which enables us to speak of *the* meaning of the word, e.g. 'table', 'son', 'member of Parliament'. This meaning may be ascertained by consulting a dictionary.

We may become clearer about the nature of meaning if we ask *how a sign* comes to be used as a *word* having meaning. We may first ask what is involved in understanding *signs*.[1] Something has already been said about this process in Chapter I. We are now concerned with signs deliberately used to signify something other than themselves, i.e. with *symbols*. Motorists are familiar with various signs giving them information about the road along which they are travelling. They may see a sign ⌈, which indicates a sharp corner ahead by resembling, to some extent, the shape of the road. The verbal sign 'bótàr ar čie' would not indicate anything to a motorist ignorant of the Irish language. But ⊣ may indicate to him 'Junction to left', if he is at all familiar with the use of these diagrammatic signs. Words do not for the most part imitate what they indicate; we have to learn that the mark SON (and the corresponding sound) stands for a male offspring, generally of a human parent. We have to learn what a *flag at half mast* signifies by observing occasions on which the flag is thus lowered. In the same way we learn a language, not by paying attention to the sound (or mark) *as a sound* (or *as a mark*), but by attending to *what* it is *some one* is using the sound (or mark) to refer to. A sign is *understood* when it is *known* what it is that some one is using the sign to signify. A hearer understands a word used by a speaker when he is referred to that which the speaker intended to indicate to him. Words, as Aristotle pointed out, are 'sounds significant by convention'. What the word is used to refer to may be conveniently called its *referend*. When the indication fails, misunderstanding results. Thus, if A (the speaker) says, 'Look at that queer thing', and B (the hearer) takes 'that queer thing' to refer to his cherished statue of Buddha, whereas A was

[1] For a further discussion, see L. S. Stebbing: *A Modern Introduction to Logic*, pp. 12 seq.

referring to a queer moth fluttering round the lamp, communication has failed. We say that B 'has misunderstood A'; he has, in fact misunderstood *what A says*.

Owing to the fact that some signs are frequently used with the same reference, dictionaries may help us to discover how a certain word is most commonly used by those who speak the language. This is so in so far as the word, whose meaning we seek to determine, is translated by a synonym which we already understand, or a description is given in terms of words already understood. We do not clearly understand a word unless we could ourselves use the word in a sentence the reference of which we understand. For this reason, good dictionaries usually give examples of sentences in which the word is used. Understanding words depends upon knowing the context.

We may take an example of learning a word which was unfamiliar. Between 1924 and 1928 a new word was used by writers on economics, 'rationalization'. We look it up in the dictionary and find the following: 'The scientific organization of industry to ensure the minimum waste of labour, the standardization of production, and the consequent maintenance of prices at a constant level.'[1] Provided that we understand the words used in this description, we now understand 'rationalization' in the context of economic statements. But we should hesitate to *apply* the word until we have had some *examples* of rationalization; we shall understand better when we realize that the word 'was coined to find a name for what was felt to be a new phase in the history of the world economic system',[2] and have discovered what exactly that 'new phase' was. Other examples of words and phrases used in new senses to refer to hitherto undescribed facts are 'complex' and 'unconscious mind' as used in modern psychology; 'inflation' and 'deflation' as used in economics.

The importance of taking note of the context in which a

[1] *Shorter Oxford English Dictionary.*
[2] G. D. H. Cole: *The Intelligent Man's Guide through World Chaos*, p. 19. It is not strictly correct to say that the word was 'coined'; it was adapted, no doubt for definite reasons of association, on the ground that to give a dog a good name is to beautify him.

5

word occurs is very great. Words, as we have seen, have a reference. Most words are descriptive, i.e. they are used to refer to characteristics or properties, which may belong to something. For example, 'bright red' is used to refer to any one of a range of colours. To understand what 'bright red' *means*, we must have actually seen something that is bright red. 'Bright red' is indefinite in its reference; hence people may not agree whether or not a given colour is to be called 'bright red' or not. There are no words which uniquely refer to shades of colours. All descriptive words are more or less indefinite. A word (or phrase) is indefinite when its reference is not uniquely determined. Clearly indefiniteness admits of degrees. We can sometimes achieve uniqueness of reference by using a combination of words in such a way that this combination *could* be used only to refer to one thing, e.g. ' the colour of the covers of this book as now seen by the speaker'; 'the present (1933) Prime Minister of this country', 'the author of *Too true to be Good*'. The significance of the word 'the' is just to indicate uniqueness of reference. In a context, uniqueness of reference may be secured by the help of demonstrative gestures, actual point-ing, bodily presentment of something, and so on. In speaking we are not *talking about* words, we are *using* them to talk *about something else*, except in the comparatively rare cases in which we are concerned only with questions of language. That is why our understanding of what is said depends upon the whole situation in which the speaker is using the words, i.e. upon the context. It is for this reason that the indefiniteness of descriptive words does not prevent us from using them in such a way that they have uniqueness of reference.

The referend of a sign used as a descriptive word is a characteristic, or a set of characteristics. When any one of a number of somewhat different characteristics could be properly referred to by the *same* word then this word is, more or less, indefinite. Some words are not only indefinite, they are also vague. A word, or phrase, is vague when it is so used that we could not tell in a given situation whether or not the word was applicable. Such words as 'bald', 'fat', 'successful business man', 'security', 'value', are vague. We cannot make a precise distinction between the state of a man's head which justifies us in calling him 'bald' and the state which would more properly be described as 'having very little hair'. Likewise with the other examples. A certain degree of vagueness in descriptive words is often quite unimportant for the ordinary purposes of life. Moreover, some words are *properly* vague since they are used to refer to a characteristic admitting of continuous variation. This is the case with 'bald' and with 'intelligent', and with all such words as 'idiot', 'imbecile', 'insane'. Intelligence can be manifested in various degrees; it is impossible to draw a sharp line between those who possess intelligence and those who do not. It is an error in good sense to insist that a speaker should draw a sharp line where, in fact, no such line can be drawn. To admit this is not to deny the important difference between being *intelligent* and being *unintelligent*; it is to admit that what *exactly* are the characteristics indicated by 'intelligent' cannot be precisely determined. An error of an opposite kind would be committed if we argued that since no sharp line can be drawn between the intelligent and the unintelligent, there is no difference between them. We shall see the importance of these considerations when we come to deal with the nature and utility of definition.

Indefiniteness and vagueness must be carefully distinguished from ambiguity. We have seen that it is not necessarily a defect in descriptive language that it should be more or less indefinite and more or less vague. The ambiguous use of words is always vicious. A word is used ambiguously when the same word is used to indicate different referends without the speakers realizing that there is a difference in what is referred to. It is only in a context that ambiguity can arise. A word considered in isolation could not properly be said to be ambiguous. Some examples may make this point clear. The sign RATIONALIZATION might be used with four different meanings: (1) in economics, to indicate what was given above on page 53; (2) in mathematics, to indicate the process of clearing from irrational quantities; (3) in modern psychology, to indicate the assigning of incorrect motives in explanation of a person's behaviour; (4) in the original use of the sign, to indicate making rational or intelligible. No doubt there are sound historical reasons why the same sign should be used with four such different meanings, and we need not in this book discuss whether or not we have not only the same *sign* but also the same *word*. The point is that 'rationalization' is *not* ambiguous. No one could be so stupid as to use it in two (or more) of these different ways without knowing that he had done so. To use a word ambiguously is *to be confused* with regard to its different indications. It is very easy to be thus confused. Ambiguity is prevalent because our thinking is so unclear. One of the most important tasks of the practical logician is to try to point out some of the various ways in which we fail to think clearly because we have failed to notice a shift in the reference of the words we use. It would take a whole volume, much larger than this little book, to deal at all properly with this topic. All that is possible here is to select a single set of words often used ambiguously, in the hope that the reader may then be led to notice other examples for himself. We need not waste time pointing out the difference of reference indicated by 'a bald head' from that indicated by 'a bald statement', nor the difference between 'a fair bargain' and 'a fair complexion', nor between 'a general strike' and 'a strike on the head'.

It is in discussions concerning politics, economics, religion, education, and art that ambiguity is most prevalent and most harmful. This is only to be expected. Where the subject is complicated, our thinking is likely to be confused; where our ordinary and passionate interests are concerned, we are likely to accept without much scrutiny any argument defending a position we want to hold. In such cases we may fail to notice a shift in meaning of the words used. Discussions concerning the General Strike of 1926 afford an amazing crop of ambiguities, as the following quotations will show:[1]

[1] These quotations (abridged) are taken from Leonard Woolf's *After the Deluge*, vol. I, pp. 304 seq. The italics are mine, and are

'Constitutional Government is being attacked. . . . Stand behind the Government . . . confident that you will co-operate in the measures they have undertaken to preserve the liberties and privileges of *the people of these islands*. The laws of England are *the people's* birthright' (*Mr. Baldwin*).

'A General Strike, such as that which it is being sought to enforce, is directly aimed at the daily life of the whole *community*' (*Lord Oxford and Asquith*).

'This General Strike was not a *strike* at all. A *strike* was perfectly lawful. . . . The decision of the Council of the Trade Union Executive to call out everybody, regardless of the contracts of those workmen they called upon, was not a lawful act at all' (*Sir John Simon*).

'The plain fact was that, not as a matter of narrow law, but as a matter of fundamental constitutional principle, when once they had a proclamation of a general strike such as this, it was not, *properly understood*, a strike at all. A strike was a strike against employers to compel employers to do something. A general strike was a strike against *the general public* to make the public, Parliament, and the Government do something' (*Sir John Simon*).

A careful examination of these statements will show how confused thinking is revealed in the shifting meanings of the italicized phrases. Are not the strikers, we may well ask, to be included among 'the people of these islands'? Do they not belong to the 'community'? Is a general strike *not* a 'strike' at all? What does Sir John Simon mean by '*properly understood*'? An examination of the speeches and writings made by various supporters of the Government, in May and June 1926, will show how uncertain was the reference intended by the words 'legal', 'war', 'enemy', as used on both sides. A conservative, for instance, tends to identify

designed to call the reader's attention to certain words which appear to me to be used ambiguously by the speakers. The reader should bear in mind that we are here concerned with an *example* of confused thinking resulting in ambiguous language; we are not concerned to take sides on the issue discussed.

'the community' with 'the middle classes', and 'the Government' with 'the State'; a trade unionist may identify 'the community' with 'the workers', and so on. We may well ask —as Prof. Laird asked, in a different connexion—'When the good of "the" community is set before us and proclaimed to be the consummation of all our loyalties, it is reasonable to ask, *What* community?'[1] To see the necessity of asking this question is to realize how easily we may be misled through a failure to recognize that a word is being used ambiguously.

A final example may be given of confused thinking resulting in a failure to recognize that language is being used both ambiguously and with an improper degree of vagueness. A writer in *The Spectator* (December 30, 1932) said:

'I do not believe in the possibility of eliminating the desire to fight from humankind because an organism without fight is dead or moribund. Life consists of tensions: there must be a balance of opposite polarities to make a personality, a nation, a world, or a cosmic system such as God planned.'

The writer gives as his reason for the conclusion that it is not possible to eliminate the human being's desire to fight—that 'an organism without fight is dead or moribund'. The word 'fight' is familiar, and the writer has failed to notice that the reference has shifted in the conclusion. In his premiss 'fight' is used in the sense of 'struggle against the environment', i.e. *tension*, in the sense of a conative urge, or drive, towards something not realized. It may be true that without a balance of 'tensions', there would be no developed personality. But in the sense in which 'fighting' means 'being at war'—which is the sense required for the conclusion—it does not follow from the fact asserted in the premiss that human beings must continue to desire to fight in order to maintain their personality. It may be doubted whether the writer had any clear conception as to what

[1] *A Study in Moral Theory*, p. 244.

exactly was the evidence upon which he was attempting to base his conclusion. Certainly his conclusion may be true, but his argument fails to support it.

It is important if we wish to think clearly to be constantly on guard to see that there is no shift of reference in the course of an argument. Logicians have been wont to insist that the middle term of a syllogism must not be ambiguous. If the middle term has one reference in one premiss, and a different reference in the other, then, although there may be only one word, there are *two* terms, and hence *no middle* term. The middle term just is that term which is *the same in both* premisses. It is the essential function of the middle term to secure that the premisses have a point of identical reference. This is the reason why the middle term must be distributed in at least one premiss. If we were thinking *only about symbols* we could secure identity of reference, and thus freedom from ambiguity, by putting the right symbol in the right place. We should thus avoid the undistributed middle, by making 'Y' appear in both premisses. But what looks the same word may not have the same reference. In the symbolic form this danger of ambiguity is concealed. Nor is it confined to the middle term of a syllogism; a term in the conclusion may fail to indicate what was indicated by the corresponding term in the premiss. This danger is both prevalent and insidious. It is so easy to attribute to words a fixity of meaning. Consider the argument: 'Of course Christians must seek peace and not war. Christians are followers of Christ, and those who follow Christ certainly seek peace.' It is not at all unlikely that the middle term of this syllogism is not used with the same reference in both premisses; possibly, also 'Christians' does not indicate in the conclusion what it was used to indicate in the premiss. It may even be the case that 'Christians are followers of Christ' may be a verbal proposition, viz. a proposition stating what a word means. If so, the conclusion merely re-states the other premiss. If not, the possibility of serious ambiguity remains. It might be replied that the speaker means '*true* Christians'. The addition of this qualification is by no means unusual. Its tendency is to beg the question. This fallacy is so prevalent that a little must be said about it here.

To beg the question is to assume the point at issue; in a (faultily) reasoned argument, the fallacy may take the form of using *as a premiss* the conclusion which the argument purports to prove. Perhaps we do not *often* commit the fallacy in the gross form in which it was committed by one of Jane Austen's characters. Unfortunately, the passage is too long to quote in full. It must suffice to quote the following: ' "Let me explain myself clearly; my idea of the case[1] is this. When a woman has too great a proportion of red in her cheeks, she must have too much colour." "But, Madam, I deny that it is possible for anyone to have too great a proportion of red in their cheeks." "What, my Love, not if they have too much colour?" '[2] Here there is no ambiguity. Both speakers probably understood the same by 'red' and by 'colour'. Yet, if so, how could the question have been so flagrantly begged? Perhaps the reader will think this discussion too stupid for ordinary life. It would not, however, be difficult to find equally glaring instances, probably in one's own reasonings, certainly in those of others. It is true that our begging of the question is usually less obvious owing to its being cloaked by unclear, ambiguous, vague language. That is why this fallacy may be fittingly dealt with in this connexion.

Let us return to the emendation of a challenged conclusion, by means of the qualification 'A *true* so and so.' For example, we make a sweeping generalization about, say, musicians and Wagner's operas, to the effect that 'No musicians nowadays admire Wagner.' When challenged, the speaker may reply, 'Well, no *true* musician does.' Pressed to make definite the distinction between 'a true musician' and just a 'musician', he might fall back, ultimately, on the test of *admiring Wagner* as differentiating the pseudo-musicians from the 'true' ones. In so doing, he would beg the question; his fallacy might be concealed from himself

---

[1] The case being whether, or not, Mrs. Watkins had too much colour.

[2] *MS. Volume the First.*

because he had no clear conception of the reference of the term 'musician'. He has used it with an improper degree of vagueness, otherwise he would have seen (we may hope) that, since—in his view—the appropriateness of calling any one a *musician* depends upon his attitude to Wagner, it would be merely verbal to say that musicians have this attitude.

No precise rules can be laid down to enable us to determine whether a given word is being used ambiguously, or with an improper degree of vagueness. There are no *principles* which could guide us in avoiding ambiguity. Only in a context is a word ambiguous. That is why symbols— such as the X, Y, Z we have used—are unambiguous; they are cut free from a context. In this abstraction from a context lies the value of symbols in revealing the formal conditions; but therein lies also their limitation from the point of view of the material conditions of reasoning. The only advice that can be offered is to be on the look-out for ambiguities. The habit of asking certain questions is a help. If we ask what must *also* be the case *if what we are saying is true*, then we may notice that what we say admits of *different* interpretations. Again, we may ask ourselves whether we are dealing with *exceptional* cases, e.g. with cases which do not quite fall under the adopted usage. It may be generally correct to say that a certain characteristic is associated with other characteristics referred to by a given word, e.g. 'religion', but it may nevertheless be incorrect in *this* case so to associate it. Thus, it might be maintained that religion is good because it involves worship. But it may be relevant to ask whether worship of *any* god, or thing, is good, or only of a god having such and such characteristics. One person's 'god' is another person's 'devil'. This point may be expressed symbolically. It may be generally true to say that X is Y, whilst in some special case that which is quite correctly called 'X', is yet not Y. An obscure perception of this divergence is often responsible for our taking refuge in such qualifications as 'a true patriot', 'a true Liberal', 'a schoolboy as such'.

The reader might suppose that the insidious danger of unclear language could be overcome if we were to define the words we use. It would, however, be a mistake to expect much help from such a practice, useful though it may be at times. Space is lacking to deal adequately with the nature of definition. It must suffice to point out that to be able to define a word is *already* to know what it signifies. In defining a word we try to set forth certain characteristics which belong to whatever the word is correctly used to refer to. Words are defined by means of other words, and ultimately by words which do not stand in need of further definition. Defining is not primarily a process of making our own thought clear; it is the signal that clarity *has been achieved*. A wiser person may help us to think more clearly by showing us how a word should be defined; but we cannot lift ourselves out of a muddle by *jumping* to a definition. For this reason the rules of definition, which it is customary for logicians to lay down, are not of practical use. These rules provide that a definition must not be too wide, nor too narrow, nor expressed in obscure or in figurative language. But the difficulty just is to know *what* would be too wide, or too narrow. For example, are we to define 'communist' in such a way that it includes all who read the *Daily Worker* and all who signed the so-called Peace Pledge and all who criticize Senator McCarthy? Or are any of these to be excluded? A logician's knowledge of formal rules affords no help.

Nevertheless, there is a stage in most argumentative discussion at which precise definition is required, whilst the search for a satisfactory definition may be itself enlightening. A good example of the need for clearly defined terms was provided in recent discussions concerning *spending* and *saving*. Broadcast talks by Mr. J. M. Keynes and Sir Josiah Stamp revealed much misunderstanding as to what exactly constitutes 'spending' and 'saving' respectively.[1] Numerous letters to *The Times* showed how wide-spread the confusion was, and how opposite conclusions appeared to be drawn from the same premisses owing to the fact that the terms employed were unclear in their reference. Mr. Keynes

[1] See *The Listener*, January 11, 1933; January 14, 1931; January 28, 1931.

argued that every pound saved put a man out of work, so that saving was not economically justifiable in a time of unemployment. Sir Josiah Stamp urged that habits of thrift were essential in a time of economic depression. He was, however, led to the conclusion that 'true saving is only another way of spending, and employs just the same'. Sir Josiah Stamp, then, distinguished between 'saving' in the sense of 'hoarding', and 'saving' in the sense of 'investing', i.e. 'spending upon a different set of objects'. This prolonged discussion would have been considerably clarified had the disputants explicitly *defined* the terms used, and distinguished between different *sorts* of saving and expenditure. The common reader is not helped in doing his·duty as a citizen when he is advised merely that 'saving' creates unemployment but 'true saving' benefits the community. The correspondence in *The Times* showed that it is easier to advise people to 'save wisely' than to show wherein lies the distinction between *wise* and *unwise* saving.[1] Clearly to understand this distinction is to know what *differentiates* one sort of saving from another sort.

Distinguishing between different sorts of the same fundamental kind constitutes *logical division*. Space permits only a few words on this topic.[2] We understand the characteristics referred to by a general term (e.g. *Saving, Liberal, Ship*) when we are able to specify its ramifications. Thus, for example, we may divide *Saving* into (1) *Hoarding*, and (2) *Investing*. We may subdivide *Investing* into (i) *Immediate Investments*, and (ii) *Delayed Investments*; (i) may be again subdivided into (*a*) *Investments aiming at the direct benefit of the investor or his descendants*, (*b*) *Investments aiming at the benefit of other members of the community*. This division is not complete, but it may suffice to show that a fuller understanding of the nature of a given class may be attained

[1] The conclusion of the Broadcast discussion between Keynes and Stamp affords a striking instance of the futility of an economic discussion in which the fundamental terms are not defined. Sir J. Stamp closed the discussion with the remark: 'In short, this saving and spending of ours are really, or ought to be, sort of sister shows.'

[2] For further discussion of Logical Division, see L. S. Stebbing: op. cit., Chap. XXII.

when we have distinguished its various sub-classes (or, as they are called, *species*), and the various sub-classes of these sub-classes. The class which is subdivided is called a *genus*, relatively to the species into which it is divided. The fundamental characteristics of the original class, or genus, must be present in each sub-class; it is these generic characteristics which justify the use of the same class-term. The specific characteristics distinguishing one sub-class from another, may justify a distinction between one species and another, e.g. between *wise* and *unwise spending*. Logical division must proceed on an orderly basis, i.e. there must be a single principle upon the basis of which one species is differentiated from a co-ordinate species. The co-ordinate species must be exhaustive of the wider class, or genus, within which they fall; otherwise, certain members of the genus will not have been included. The catalogue of a well-arranged library exhibits a logical division. As the catalogue suggests, the principle of division is relative to purpose. We may divide books according to *authorship, date of publication, subject-matter, binding*, etc. Each principle of division would yield a different arrangement of the classes concerned. What would, in given circumstances, be the most fruitful division depends upon what differentiating characteristics are most *relevant* to the question at issue. A logical division is fruitful when it gives rise to inferences relevant to the topic under discussion, i.e. when, from knowing the place of a sub-class in the orderly arrangement of classes, we can infer how its members resemble and differ from the members of other sub-classes.

The reader will have noticed that making a satisfactory logical division depends upon our knowing the material conditions, so that, here again, formal rules do not afford much practical help, except in so far as they may aid us to recognize *how* a proposed division fails to be sound. It is easy to insist upon the need for *relevance*; it is often difficult to know *what* is relevant. A wise man will not attempt to argue about a subject on which he is ill-informed. Unfortunately, many of the topics on which we hold strong opinions are topics concerning which we are sadly ignorant. The ignorance of relevant considerations leads us to construct faulty

arguments and renders us a prey to the unscrupulous disputant. A common fallacy of this type is that of *the irrelevant
conclusion*,[1] which consists in establishing a conclusion which
is other than the conclusion intended to be proved. For
example, a disputant may attempt to throw doubt upon an
opponent's statement by asserting that it is to the advantage
of the opponent to believe in the truth of his statement.
This, however, is not the point at issue; hence, the argument is irrelevant *unless* it can be shown that the opponent's
sole reason for accepting the statement is his desire for it
to be true. It is by no means uncommon for conservatives
to argue that socialism must be unsatisfactory since it is
based upon the envy of the 'have-nots' for those who have;
socialists, on the other hand, sometimes seem to suppose
that it is a sufficient argument against capitalism, to show
that capitalists desire to keep what they possess. Another
form of irrelevant argument consists in blackening the
characters of those who support a proposition; still another
form is found in ridiculing the supporters. A joke often
provides an excellent diversion, but it is a *diversion*, a turning
away from the point at issue. The forms which irrelevant
arguments may take are too numerous to be dealt with here.
One more, very common, form may be noted. In a recent
trial, the counsel in defence of a convicted prisoner, sought
to mitigate his sentence by calling the judge's attention to
the fact that the prisoner had a wife and five children. This
is clearly an irrelevant argument. If, however, the counsel
were to plead that the prisoner's previous record had been
good, and that he had, apart from this lapse, dealt honour-
ably in business, then his argument would have been relevant. The *form* of an irrelevant argument is: You must
accept Q because you accept P, where, in fact, P does not
establish Q. Thus formally stated it might be supposed that
no honest thinker could be so misled. The difficulty is,
however, that our arguments are not set out briefly, in clear
language, and consequently we easily fail to perceive the
want of connexion between P and Q. The use of emotionally

[1] This is known as the fallacy of *ignoratio elenchi* (i.e. ignoring
the point at issue).

charged language may create an attitude of mind which makes us accept an unsound argument. Thus we find socialists accusing capitalists of 'robbing' the poor; we find capitalists dubbing unemployment insurance as 'the dole'; we think of our enemies (in war) as 'murderous foes', and of our own men as 'heroes'. The reader will be able to supply other examples. The only way to avoid being led into unclear thinking of this kind is to attempt to translate language directly arousing emotional attitudes into non-emotive language, and to consider whether the reasons urged against one's opponent would be relevant against oneself. We are all inclined to the fallacy of 'special pleading', i.e. accepting (or refusing) in one's own case an argument which one refuses (or accepts) on the other side. For example, a person may condemn the 'dole' on the ground that the recipient has not earned it by work, whilst accepting the view that those who inherit wealth may live on an unearned income.

A closely allied fallacy consists in asserting an indisputable contention and thence proceeding to another proposition in no way related to the former. The hearer accepts the platitude, and may fail to notice that the contention in dispute is in no way established. For example, it may be argued that 'human beings are governed by primary human instincts, not by socialist theories', and that *therefore* socialist theories are wrong. This *therefore* is a *non sequitur*. Socialist theories may be profoundly mistaken, but it is not in such a way that they can be disproved.

It is useful to cultivate the habit of asking oneself whether a given statement is supported by the argument offered. If so, the premisses must not only be consistent with the conclusion but must provide some *reason* for it. This reason will, we have seen, be valid in any other argument of the same form. The language used must be free from ambiguity; the point at issue must be definite. We do not *disprove* a proposition by showing that the argument offered in its support is unsound, but, unless we are offered another, and a sound, argument in support of it, we have no reason for accepting it as true.

CHAPTER V

THE ESTIMATION OF EVIDENCE

'A straw will show which way the wind blows.'—OLD SAW

THE connexion between a mass of evidence and that which it evinces may be approached from two different points of view. The first is exemplified in the attitude of legal counsel, for the defence, or for the prosecution; the second, in the attitude of a detective attempting to discover the man who did the deed. For brevity, we may refer to the mass of evidence as the *data*, and to that which it evinces as the *probandum*. The counsel accepts the *probandum* as already determined; his problem is to select from miscellaneous, and possibly conflicting data, just those facts which point to the already accepted *probandum*. The detective seeks a *probandum* which is, at the outset, completely undetermined; his problem is to determine the *probandum* by examining the data, selecting what is relevant, and recognizing its significance. His selection is guided by an hypothesis, more or less capable of explicit formulation. His thinking involves the three steps mentioned in Chapter I; if the conditions constituting the problem are at all complicated, he may need to try out several hypotheses before he is satisfied that he has hit upon the correct solution. The detective's task is more difficult than that of the counsel. The data may point in many different directions; at first sight it may even be the case that *no* definite *probandum* appears to be indicated by the available data. Once the case is completed, the detective, no less than the counsel, may present his conclusions in deductive form. Nevertheless, his reasoning remains essentially inductive. For example, the detective may argue: 'A's boots fit these footprints in the flower-bed; therefore, A made these footprints.' The cogency of this argument would depend upon certain assumptions, e.g. that one person's boots are discernibly different in shape from any one else's, and that footprints in flower-beds can reveal the differences between the boots that caused them and all other boots. When the assumptions are made explicit the generalizations involved therein are obviously far from being certain. A detective's argument will be much stronger when it is based on the generalization that no two people have the same finger-prints. The warrant for this assumption is to be found in the *observed* fact that every person tested for finger-prints is found to have peculiar markings. It should not be necessary to multiply illustrations of the contention that the significance of the observed facts—constituting the original data—is wholly due to our knowledge of the regular ways in which one happening is connected with other happenings. The cogency of the counsel's argument depends likewise upon the previous acceptance of premisses obtained by inductive generalization.

It is customary to distinguish three modes of inductive inference, viz. *analogy, generalization, circumstantial evidence*. Although the distinction is useful up to a point, yet the three modes are of fundamentally the same nature. Each of them is based upon the recognition of relevant resemblances and relevant differences. Inference by analogy consists in inferring that, since two cases are alike in certain respects, they will also be alike in some other respect. For example, since Mars resembles the Earth in certain respects, we infer that Mars also is inhabited. This may be a very risky inference, for Mars differs from the Earth in some respects, and these differences may be relevant to the property of being inhabited. If so, then whatever may be the extent of the resemblance between Mars and the Earth, this resemblance is unimportant from the point of view of the given inference. Any respect in which Mars resembles the Earth (e.g. *revolving round the Sun*) puts Mars into a class consisting of at least two members, viz. *Mars and the Earth*. This resemblance may then be the basis of a generalization.[1] Since the members of *any* class resemble each other in some respect and differ in others, the argument from resemblance must be controlled. Hence, we are led to distinguish

[1] See pp. 14–15 above.

between *essential* (or important) and *unessential* (or unimportant) resemblances and differences, and thus to form *classes*. Generalizations relate to classes, and are thus based upon analogy. We resort to simple analogy (resemblance between individual instances) only when the circumstances are too complex, or the case too rare or too unfamiliar, for us to be able to fall back upon the generalization involved in the recognition of a class.

The phrase 'circumstantial evidence' is most usually employed to designate the form of reasoning in which a set of evidentiary facts cumulatively point to a certain definite conclusion although no single fact itself suffices to indicate that conclusion. This is the form of reasoning employed by detectives—at least in detective novels. Poe's *Rue Morgue* affords a good example; the committee of investigation—discussed in Chapter I—provides another example. The distinguishing characteristic of this mode of inference lies. in the cumulative force of a set of facts taken together. It would be a mistake to suppose that this mode of inference is confined to criminal investigation. On the contrary, all reasoning of the form—if F_1 and F_2 and F_3 and . . . , then *very probably P; but* F_1 *and* F_2 *and* F_3 *and* . . . , *therefore very probably P*—falls under this mode.

It should be observed that inference from circumstantial evidence involves generalization, and therefore analogy. We saw this to be so in the case of arguing about footprints. To infer that F_1 makes P more or less likely is to rely upon certain *general* characteristics of F_1. In short, the significance of each separate fact—i.e. each separate item in the evidence—depends upon the thinker's knowledge that things of one kind are associated *more or less frequently* with things of another kind. For example, when we infer that A probably made these footprints, we might be relying on our knowledge that A wears size 11 boots and that people who do this form a very small proportion of the population. And if we argue 'A's footprints were made outside a window at night, so A probably visited the house with criminal intent', we are basing our inference on knowledge of the usual (the most frequent) behaviour of human beings;

though plainly we would not be justified in assuming that *all* people who approach the window of a house at night intend to commit a crime. An inference from circumstantial evidence is not deductive. The conclusion is not that the set of facts *entails* P; it is that they make P probable. Hence P may be false, notwithstanding that the facts are as reported. The inference accordingly, is inductive.[1]

Inference from circumstantial evidence is often regarded as a chain argument. But if a chain is not stronger than its weakest link, then this description is inept. A single fact, F_1, may weakly suggest P, yet the strength of the cumulative evidence may be considerable. Its strength is due to the consideration that P fits all the facts and that there is no reasonable alternative to P which fits all the facts. The weapon with which the murder was done may belong to A, and A may have had a motive to commit the murder, and may have had the opportunity, and yet the murderer might be B. It is certainly true that in real life, as well as in detective stories, it may occasionally happen that a person is entangled in a web of circumstantial evidence pointing to the conclusion that he has committed a crime of which he is, none the less, innocent. Even assuming that the difficulties of ascertaining the relevant facts and of obtaining reliable evidence from eye-witnesses have been overcome,[2] we are forced to admit that circumstantial evidence cannot suffice to yield a *certain* conclusion. In inductive inference we are never in a position to maintain that *no other conclusion* is consistent with the evidence. But if all the facts point to P, and no alternative possibilities are discovered, then we feel it would be stretching the 'long arm of coincidence' too far, to reject P on the ground that some unthought-of alternative would explain away the set of facts which together indicate P. A proper discussion of this topic would take more space than is at our disposal. It must suffice to point out that the inference is more reliable in proportion as (1) each of the accepted facts is adequately explained by

[1] See p. 15 above.
[2] On the difficulty of obtaining such reliable evidence, see A. W. P. Wolters: *The Evidence of our Senses*, Chap. IV.

P; (2) it is unlikely that any relevant facts have been over-looked; (3) it is likely that if there had been contradictory facts they would have been noted. These are big provisos. Nevertheless, we may sometimes have reasonable confidence that the facts warrant the conclusion.

How, we may ask, is this reasonable confidence secured? We have to rely upon our knowledge of inductive generalizations. These are of two kinds: (i) concerning *uniformities* in the behaviour of things, and such generalizations can be expressed in the form *All X's are Y's*; (ii) concerning the *relative frequency* with which things of a given kind are associated with things of another kind, and these may be expressed in the form *Such and such a proportion of X's are Y's*. As we have seen, in discussing circumstantial evidence, we often have to employ generalizations of kind (ii). Scientists too must often use such generalizations as pre-misses, e.g. in meteorology and applied psychology. The disadvantage of using such generalizations, in trying to solve practical problems, is that the conclusions of our inferences cannot be certain, but only more or less probable. Scientists will always try to replace generalizations of kind (ii) with those of kind (i).

One important kind of generalization which scientists seek to establish concerns the causal properties of things, i.e. the ways in which a thing always behaves in relation to other things. We shall now discuss certain principles implicit in the experimental methods which scientists use for this purpose. These principles are derived from the funda-mental notion of causation. We shall define this notion by saying that two events, X and Y, are causally connected when X is both a necessary and a sufficient condition of the occurrence of Y. In other words 'X causes Y' implies both that *All events of kind Y are accompanied or preceded by an event of kind X* and that *All events of kind X are accom-panied or followed by an event of kind Y*. X is said to be the *cause*; Y is said to be the *effect*. Two principles follow directly from the nature of a cause, i.e. (1) Nothing is the cause of an effect which is absent when the effect occurs; (2) Nothing is the cause of an effect which is present when the effect fails to occur. Accordingly, in seeking for the cause of an occurrence, Y, we shall look for situations in which Y is present, and for situations resembling the former in many respects but differing from them in the absence of Y. These principles yield two derivative principles, which may be called, respectively, the *Principle of Agreement* and the *Principle of Difference*. Two examples may suffice to show how these principles are used.

A certain man finds that on eight successive Tuesdays he has a headache; but on no other days during those weeks has he had a headache. He asks what has happened on the Tuesdays which has not happened on other days. He remembers that on each of those days he has returned from the City by the Underground Railway, whereas it is his usual custom to return by bus. But on those Tuesdays he had an early after-dinner engagement to play chess at a friend's house, and to get there in time he had to travel by the quicker route. On other evenings he plays chess in his own house, sometimes with this friend, sometimes with others; he does not then need to be home earlier. On these days he does not have a headache The journey by Underground is common to all Tuesdays and nothing else seems to be both common and peculiar to the days on which he gets a headache. He therefore concludes that the journey in the Underground is causally connected with his headache. In reaching this conclusion he is employing the Principle of Agreement. The conclusion is by no means certain. Yet, if he has played chess with the same friend on other occasions when he did not have a headache, and on these occasions he had travelled home by bus, then the Principle of Agreement makes it reasonable to suppose that the Underground journey is responsible. If, further, he came home by Underground one day and had a headache thereafter although he did not go to his friend's house, nor play chess, then the probability that the conclusion is correct is strengthened, since the cause must be present when the effect is.

A healthy man eats a liqueur chocolate. Almost immediately he falls down dead. It is concluded that he was poisoned by what he had just swallowed. This conclusion is reached by application of the Principle of Difference. At one moment the man is alive and well; a few moments later he is dead. Nothing appears to have happened *except the eating of the chocolate*; hence, no other factors can be responsible. If it is then found that cyanide of potassium had been put into the chocolate, then we shall be confident that this poison caused his death. No doubt we should then reason deductively as follows: Whoever swallows a certain amount of cyanide of potassium dies immediately; this man has swallowed such an amount of cyanide of potassium; therefore he dies. The reader should observe, however, that no one would examine the chocolates to see if they were poisoned *unless* it had been assumed that the eating of the chocolate were causally connected with *this* man's death. Many people eat chocolate and continue to live. Why, then, should the chocolate have been examined? The reason is that the eating of the chocolate was the sole new factor introduced into the situation in which the man had been alive and healthy. Most people now know that cyanide of potassium is poisonous, and that chocolate is not. But at one time it was a discovery that this property belonged to cyanide of potassium. This discovery could be made only by noticing what happened when cyanide of potassium was absorbed by a living organism. If no other factor in the situation had been changed, then the Principles of Causation justify us in concluding that this additional factor was the cause of the observed effect. The condition that *only one factor varied* is of great importance. Neglect of it is partly responsible for the very common fallacy of *post hoc ergo propter hoc*, i.e. the fallacy of concluding that what has immediately preceded an occurrence is the cause of that occurrence. For example, a man curses his enemy, who shortly afterwards dies; there is a 'change in the moon' and then a change in the weather. To argue that the second (in either case) is causally dependent upon the first is to mistake a temporal conjunction for a causal connexion. We are tempted to fall into this fallacy when one or other of the two occurrences is especially striking. We cannot even argue from a *constant* conjunction of two occurrences to a causal connexion; we require to observe a situation in which *one* factor can be eliminated. The fallacy of *post hoc ergo propter hoc* is responsible for many popular superstitions. The man who trusts to his mascot to help him win a match may never have tried what would happen if he left it at home.

The two examples previously given should suffice to show that the discovery of causal connexions depends upon an analysis of a complex situation. Certain features of the situation must be simply judged to be irrelevant.[1] The man who had a headache on Tuesdays would judge *Tuesday* as such to be irrelevant; the day of the week is important only in relation to how its occupations differ from those of other days. The colour of the carpet upon which the poisoned man was standing would also be judged irrelevant, since, presumably, he and other people had stood on it before without ill-effect. It is, however, easy to rule out as irrelevant factors which are indispensable. For example, it was often assumed that the colour of the walls of a sick-room had no effect upon the condition of the patient. It is now known that certain mental patients are made worse by seeing some colours, and are aided by seeing others. The only way to avoid making mistakes of this kind is to resort to comparison of cases in which different factors are varied. The most satisfactory procedure is to test by experiment, i.e. by deliberately varying a given factor and observing what happens. In an experiment the observer is able to control the conditions in such a way that he can vary the factor he is investigating without thereby varying other factors. Wherever experiment is possible, hypotheses with regard to possible causes can be tested. It is not difficult to see that the field for experimental testing is limited to those situations in which the observer can deliberately arrange to initiate those changes the results of which he wishes to observe. Just as a skilful barrister, in cross-examination of a witness, asks those questions which are most likely to yield the answers he wants, so a skilful experimenter arranges those conditions the observation of which will answer the questions constituting his problem.

[1] Cf. pp. 2–3 above.

Much· might be said about the technique of experiment. But to do so would require another small volume. For our purposes, however, it is not important to stress the part played by experimental investigation in the more advanced sciences. Nor need we pause to consider the bearing of experiment upon quantitative investigation. From the strictly logical point of view, the most complicated scientific experiment reveals only the same logical principles as are exemplified in our ordinary reasonings concerning matters of fact. In both alike what matters is that we cannot formulate a question save on the basis of previous knowledge; we must make judgments of irrelevance, since no situation presents *only* those features which are significant for our problem; we must analyse the situation under investigation in order to discover its relevant likenesses to, and differences from, other situations of the same general nature. Those principles which control sound generalizations concerning classes are also the principles which lie at the basis of causal investigation.

CHAPTER VI

THE GROUNDS OF OUR BELIEFS [1]

'It is undesirable to believe a proposition when there is
no ground whatever for supposing it true.'

BERTRAND RUSSELL

WE all commonly entertain many beliefs for which we have little, or no, evidence. Some of these beliefs may be baseless, but some may be capable of being supported by sound evidence, which we could discover if we wished. Frequently, however, we do not know, and have never thought to inquire, what this evidence is. When, however, a cherished belief is challenged we may be moved to argue in its support; when a doubt has occurred to ourself we may seek to remove that doubt. In seeking to resolve a doubt we are seeking premises from which the proposition in question follows, or which can at least be adduced as affording *some* evidence for its truth. These premises provide logical reasons justifying belief in a given conclusion. Frequently it happens that the evidence is not sufficient to *imply* the conclusion whilst it is sufficient to justify the belief that the conclusion is probably true. To say that the truth of a proposition is more or less probable is to say that there is *some* evidence (more or less strong) in its favour and no *conclusive* evidence *against* it. To have conclusive evidence *against* a proposition is to have a logical reason for *disbelieving* it. The notion of believing *that so-and-so is probably true* is familiar to common sense. If some one said, 'In my opinion *there will be another great European War before 1940*', he would be tacitly admitting that the italicized statement is not known to be certainly true, whilst asserting his belief that the available evidence renders its truth probable. The reader will have no difficulty in understanding this notion of probability. He should, however, observe that the

[1] The word 'belief' is used throughout simply as short for 'that which is believed'.

words 'probability', 'opinion', and 'belief', are not used with precision in ordinary conversation. We sometimes assert an opinion when we have no evidence at all in favour of that opinion. In such a case it is incorrect to say that the proposition opined is *probably true*, for to say this is just to say that there is *some* evidence in its favour. Whilst, therefore, some of our beliefs may not be based upon evidence, we are not entitled to claim that a given proposition is probably true, unless we can give some reasons for accepting it. Probability admits of degrees, varying between the two extremes of *certainly true* and *certainly false*. Thus probability is not equivalent to mere possibility, nor improbability to impossibility. Probability is relative to the evidence, so that the truth of a proposition may be extremely probable (or improbable), in face of the available evidence, and may yet be false (or true). For example, the evidence now available with regard to the statement *that there will shortly be a European War* may make its truth very probable, and yet the statement may be false. Some change, not now foreseeable, in the attitude of nations might occur. But, since *unforeseeable* occurrences are necessarily *unforeseen*, this mere possibility is not *evidence*. It is unreasonable to entertain a strong degree of doubt with regard to a proposition which has been shown to have evidence rendering its truth probable.

Believing must be distinguished from having knowledge. Beliefs may be false, but what is known cannot be false, since 'false knowledge' is a contradiction in terms. Further, we may have a true belief where we do not have knowledge, for we may entertain a belief, which is in fact true, only because we believe something else which is false. For example, a juror may truly believe that an accused prisoner is innocent simply because the juror has taken a dislike to the principal witness and refuses to believe he is speaking the truth, whereas his testimony may be correct. The juror could not then be said to *know* that the accused prisoner is innocent even though his belief were true. A judgment which is accepted on false grounds cannot be *known* to be true, even when it is in fact true, so that in believing it we should be believing truly. We should only be in the position of happening to believe *what is the case* without knowing *why it was the case*. Most of what commonly passes for *knowledge* is at best only opinion or belief having a considerable degree of probability of truth. From the practical standpoint, however, it would be inconvenient to refuse to accept as knowledge what has, in fact, only a high degree of probability. A high degree of probability is often called 'practical certainty'. A reasonable man should not refrain from acting upon a practical certainty as though it were *known* to be true. In England, for instance, it is customary for a judge, at the trial of a person accused of murder, to instruct the jury that an adverse verdict need not be based upon the belief that the guilt of the prisoner has been 'proved', but upon the belief that the guilt has been established 'beyond reasonable doubt'. To be 'beyond reasonable doubt' is to have sufficient evidence to make the proposition in question so much more likely to be true than to be false that we should be prepared to act upon the supposition of its truth. Many of our most important actions have to be performed in accordance with beliefs of such a kind. A healthy youth acts reasonably if he prepares himself for a career, notwithstanding the possibility that he may die before its fruition. If, however, he sets out on an extremely hazardous adventure, he would act reasonably in making his will beforehand.

It is important that our beliefs should not be such as a reasonable man would be compelled to reject. Nevertheless, we often have to act upon a definite belief although there is much to be said on the opposite side. This is the case with many beliefs about politics, about our educational policy, about our charities. When we *must* act in one way or the other, it is simply stupid to refrain from committing ourselves to a definite belief, even though we may see clearly what may be urged on the other side. All that we can do is to act in accordance with that belief which seems to us, after due thought, to be more likely to be true than is the contrary belief. The recognition that the case of the other side is not negligible may well make us more tolerant, but it should not render us merely undecided in action.

Sometimes our beliefs are erroneous because we have accepted an unrestricted generalization when only a restricted one would be justified. Thus we may hold that Frenchmen are always clear thinkers, or that Englishmen are always honourable, or that people who speak fluently do not think profoundly; in each case it might be that the substitution of 'usually' or 'very often' for 'always' would render the belief justifiable. In some sciences much use is made of a form of statement which enables us to substitute, for such generalizations as the above, a more precise proposition which has a better chance of truth. An example may make the point clear. Let us consider the unrestricted statement, *Fluent speakers are not profound thinkers*. As it stands this suggests either that there is some causal connexion between the ability to speak fluently and lack of ability to think profoundly, or that the two characteristics happen to be conjoined. On either alternative we should be ready to regard fluent speaking as a *sign* of superficial thinking. It would be reasonable, however, to ask whether there is a greater proportion of superficial thinkers among those who speak fluently than among those who speak hesitatingly or slowly. We must, then, consider the four classes: (i) fluent speakers; (ii) slow speakers; (iii) superficial thinkers; (iv) profound thinkers. Let (i) be represented by X, (ii) by non-X, (iii) by non-Y, (iv) by Y. Our problem is to discover whether the proportion of X's which are non-Y exceeds the proportion of non-X's which are non-Y. To solve this problem we must carry out a statistical investigation, i.e. we must examine a number of speakers, taken at random, divide them into the four groups, XY, X non-Y, non-XY, non-X non-Y, and then determine whether a substantially higher percentage of the X group than of the non-X group fall into the non-Y class.[1] If so, then we should be justified in saying that fluent speakers *tend* not to think profoundly. This statement would justify us in believing that a fluent

[1] It is not possible here to do more than suggest the value of precise statistical investigations. I do not wish to imply that numerical ratios could be profitably introduced into ordinary discussion, but merely to call attention to the wisdom of refraining from sweeping generalizations which have not been tested.

speaker is more likely than not to think superficially; but it would not justify us in feeling certain that fluency of speech must be combined with lack of profundity in any particular case.

We require knowledge, or at least true belief, for the ordinary purposes of life. A belief is justified when adequate evidence is adduced in its support. Some of our beliefs, however, stand in no need of justification, since they have consequences but no grounds. Such beliefs may be called *underived beliefs*. They are to be contrasted with *derived beliefs*, i.e. beliefs *capable* of being based upon evidence. There appear to be two kinds of underived beliefs: (1) beliefs concerning sense-experience and memory; (2) pre-reflective beliefs of common sense. The ordinary man does not question his belief, for instance, that he is alive, or that he feels tired, or that he hears a loud noise. He sees the difference between asserting 'I hear a loud noise' and 'I hear a pistol-shot'. He would not regard it as reasonable to question whether he hears a loud noise; he would say he *knows* he does. But he might admit the possibility that what he heard was not a pistol-shot but the sound of a bursting tyre. He might admit that the assertion *That was a pistol-shot* could have been derived from the two premisses: *That was a noise of a certain kind* and *Only pistol-shots make that kind of noise*. A belief which could significantly be questioned *could* also (if true) be derived even if no one had in fact ever derived it. A belief which was not only underived but also *underivable* could not significantly be questioned, since it would be meaningless to ask what were the grounds of a belief which could have *no* grounds. The second kind of underived beliefs may be called 'intuitive beliefs', provided that we remember that intuitions may be mistaken. Examples of intuitive beliefs will be found in the various logical principles we have mentioned.

There are also two kinds of derived beliefs: (i) those derived from what other people tell us, i.e. from testimony; (ii) those derived by inference from (1), (2), or (i). These divisions cannot be sharply maintained since (i) could be reduced to (ii). Possibly many intuitive beliefs could be derived by inference. Usually, however, we do not so derive them. This fourfold division of kinds of beliefs may. be conveniently adopted here.

What we can know directly by means of sense-experience and memory constitutes a very small portion of our unquestioned beliefs. This little store of knowledge is considerably increased by accepting testimony and by deliberate drawing of inferences from what is thus known. Inference is indeed the most common way of increasing our knowledge. It is not to be suggested that testimony should be unhesitatingly accepted, nor that our pre-reflective beliefs should never be questioned. But doubts are fruitful only when we are prepared to think simply in order to discover whether our beliefs are justifiable, and when we have some knowledge of how to set about justifying them.

It is this problem of justification which interests the logician. The psychologist is interested in the analysis of mental attitudes and in the problem *how* we come to believe or doubt something. As practical logicians we are interested in the latter problem only in so far as knowing *how* we reach the beliefs we entertain would help us in practice not to believe or to doubt without justification. We want to be able to distinguish between *good* reasons and *bad* reasons, i.e. to distinguish between an argument which is logically sound and one which, although it convinced us, was nevertheless unsound.

There are at least five different ways in which we may come to hold some non-intuitive belief. (1) We may believe a proposition because we have frequently heard it asserted and have never thought of questioning it. We may even be unaware that our acceptance is based upon what people have told us, for we may have grown up with the belief. The accepted commonplaces of thought fall under this head, e.g. *Murder will out.* Many beliefs accepted in this way are true, but some are not; if we do not recognize that a belief thus reached may, *for all we know*, be erroneous, we may some day get a severe shock. (2) We may accept a belief on the authority of a parent, or a teacher, or a church, or some social institution. Such acceptance presupposes the belief that the authority is reliable. This, again, may well be the case, but mere reliance on authority involves risk of error. (3) Our belief in a given statement may be based upon acceptance of the testimony of an expert. The field of knowledge is so extensive that no one can hope to have first-hand knowledge concerning many interesting and important topics. An expert is a person who has made a special study of a given subject and has thus acquired competence therein. This competence renders him reliable within the field of his study. It would be foolish for a layman to question the correctness of a scientific statement made by an expert in that branch of science. For instance, a person who has never studied the marriage customs of primitive peoples has no good ground for believing that polyandry is contrary to human nature if it is the case that anthropologists have professed to provide evidence that some tribes practise polyandry. It is true that experts do not always agree, but their disagreements cannot be evaluated by a layman . Moreover, sometimes they do agree. It is well to remember, as Bertrand Russell has pointed out, 'that when the experts are agreed, the opposite opinion cannot be held to be certain', and 'that when they are not agreed, no opinion can be regarded as certain by a non-expert'. The reader may think that no one would dissent from this counsel. Nevertheless, we all do tend to hold firmly certain views about matters arousing our emotional interests, although these views would be decisively rejected by all the experts. Whilst it is reasonable to accept expert testimony, it is foolish to allow an expert in a special subject to dictate to us outside the limits of that subject. At the present time there is a tendency to allow scientists to tell us what we ought to think about subjects in which they have no special competence.

The three cases just discussed relate to beliefs which are derivative, but which we have not ourselves reached by explicit inference. The next two cases relate to beliefs consciously accepted after a process of questioning. We are now concerned with ways of resolving doubt or of removing admitted ignorance. Whenever we strongly entertain a belief which we consider important, we are tempted to induce other people to share our belief. The desire to secure agreement may be so strong that we may be willing to use any means capable of attaining our aim. Cases (4) and (5) fall under this heading. They are to be distinguished by the nature of the means employed. One way may be called the method of persuasion, the other the method of conviction.

It is often said that the art of persuading others to agree with us is the art of oratory. But it is to be feared that many possess the power to persuade who do not possess the attractive gift of oratory. Certain wiles, however, are possessed in common by orators and by those who make public speeches. An illustration may make clear the method of persuasion. In October 1932, Mr. Stanley Baldwin broadcast a speech, shortly after the resignation of certain cabinet ministers from the National Government. Mr. Baldwin stated that it was his intention to give the reasons why he and his friends intended to 'stick to the Government'. Without further preamble he said:

'A little over a year ago the ship of state was heading for the rocks. The skipper had to change his course, suddenly, and many of his officers and most of his crew deserted. It was a case of all hands to the pumps, and I signed on with my friends, not for six months or a year; I signed on for the duration, be the weather fair or foul, and I am going to stick to the ship, whether it goes to the bottom or gets into port, and I think the latter end is a good deal more likely.'

This is a skilful statement. It at once induces an attitude of acceptance. Every one would admit that it is dastardly to desert a ship heading for the rocks, could working at the pumps save it. Few of his hearers would pause to ask whether the skipper and the rest of the crew might have escaped the rocks had they joined the deserters—probably in the boats. A British audience in 1932 would have an emotional attitude to the phrase 'signed on for the duration', since it awakens memories of sacrifices made during the recent war. A wave of sympathy would be felt for the brave officer who will 'stick to the ship' even unto death. But the whole force of the argument, *if it be intended to provide good reasons* for remaining in the National Government, depends upon the soundness of the comparison between the position of a government, on one hand, and the position of the officers and crew of a ship, on the other; and between the position of the National Government in 1931 and a ship on the rocks. No training in logic is required to enable us to see that in no single detail is there any relevant likeness between the things compared. It might as well be retorted that the brave officers were those who first showed the way by springing into the angry sea. This argument may persuade a stupid electorate; it cannot convince any one.

The method of conviction consists in seeking to secure acceptance of a proposition by showing that it is derived from sound reasons, i.e. from evidence adequate for its support. It has been the aim of this book to indicate the conditions of adequate evidence. Lack of space, as well as the incompetence of the author, has made a full treatment impossible. Enough has been said if the reader has been convinced of the importance and the difficulty of clear thinking with regard to the grounds of our beliefs. A pedantic demand for the grounds of all our beliefs is not to be encouraged. But if we do not wish to be at the mercy of a skilful but unscrupulous persuader, we must have some awareness of what *sort* of argument can properly be adduced for any statement we are asked to accept. It is only possible here to notice that different sorts of statements require different sorts of evidence. In the case of mathematical statements, the evidence offered is abstract, and the form of argument must be rigidly deductive. Statements in physical science can never be supported by purely deductive reasoning, since their ultimate test is to be found in experimental confirmation. Nevertheless, deductive reasoning plays a considerable part. In the case of the social sciences, such as economics, sociology, and anthropology, as in political science, statements should be supported by evidence obtained from observation, aided by statistical investigation. There is no place for pre-reflective beliefs. It is foolish to believe that socialism is incompatible with human nature *unless* we can state definitely with what established psychological generalizations the theory of socialism comes into conflict. In this statement 'capitalism' could equally well be substituted for 'socialism'.

Finally, our beliefs about matters requiring expert investigation must be accepted on the authority of the experts. By itself, a training in logic does not make us competent to have opinions about physics or about religion. In one sense we do not have *beliefs* about subjects of which we understand nothing; we have only opinions capable of verbal expression. To be able to say 'I accept the fact that nothing can travel faster than light' is quite different from the *insight* that this is so. Insight comes from clear apprehension of relevant connexions.

There is only one reliable method by which we may increase our knowledge, or, where knowledge is impossible, be led to entertain reasonable beliefs. The method consists in apprehending the relevance of that which we already know to that which we do not yet know but are able to discover because what we already know is significant of that which is unknown.

REFERENCES FOR READING

IN addition to the works mentioned in the text, the following books are recommended to those who desire to read more on some of the topics which have been discussed:

R. H. THOULESS: *Straight and Crooked Thinking*. (Hodder & Stoughton.)

L. J. RUSSELL: *An Introduction to Logic. From the Standpoint of Education*. (Macmillan & Co.)

A. SIDGWICK: *The Use of Words in Reasoning*. (A. & C. Black.)

E. A. BURTT: *Principles and Problems of Right Thinking*. (Harper & Brothers.)

L. SUSAN STEBBING: *A Modern Elementary Logic*. (Methuen & Co.)

INDEX

About, two senses of, 34
Abstraction and analysis, 13
Affirmation and denial, 28–30
Affirmative proposition, 29
Agreement, Principle of, 72
Alternative argument, 47
    proposition, 45
Ambiguity, 55 seq.
Analogy, 68, 69
Antecedent, 44
Applicative Principle, 38, 50

Belief, 77 seq.
    derived and underived, 80

Causation, fundamental principles of, 71 seq.
Class:
    and class-name, 11, 13
    and things of a certain sort, 10, 31
    grouping into a, 11 seq., 68
Compatibility and incompatibility, 26
Compound proposition, 44
Conjunction, 44, 45, 73
Consequent, 45
Contradiction, 28, 30
Contrary, 28
Conversion, 34 seq.
Conviction, method of, 84

Datum:
    and conclusion, 6
    and premiss, 15
Deduction, Principle of, 19.
    See Inference
Definition, 62 seq.
Dictum de omni, 39, 41, 50
Difference, Principle of, 72
Dilemma, 48
Disjunctive proposition, 45
Distribution, 34, 35, 39

Division, Logical, 63 seq.

Empirical generalization, 15
Entailing, 17, 18, 27, 40
Equivalence, 27, 34 seq.
Evidence:
    and conclusion, 6
    and probability, 77 seq.
    circumstantial, 68 seq.
Excluded Middle, Principle of, 31, 50
Excluding from a class, Principle of, 38, 41
Experiment, 74 seq.

Fallacy:
    of begging the question, 60
    of Consequent, 47
    of Irrelevant conclusion, 65
    of *Post hoc ergo propter hoc*, 73
    of Special Pleading, 66
Form, of an argument, 22. *See* Reasoning

Generalization, 10, 15, 16, 32 seq., 68 seq.
General property, 15

Hypothesis, and thinking, 4 seq., 67 seq.
Hypothetical argument, 46
    proposition, 44

Important:
    and purpose, 12 seq.
    defined, 12
    property, 14
Inclusion, relation of, 42
Indefiniteness, and ambiguity, 55 seq.
Independent propositions, 18 seq.

89

Milton Keynes UK
Ingram Content Group UK Ltd.
UKHW022357061024
449327UK00031B/2554